Thermal Environment

J. N. CHALKLEY ARIBA
H. R. CATER MIHVE, AMBIM, MRSH

THERMAL ENVIRONMENT

FOR THE STUDENT OF ARCHITECTURE

THE ARCHITECTURAL PRESS: LONDON

Preface

This book is in itself only an introduction to the field of Environmental Engineering and the associated practices of Heating, Ventilating and Air-Conditioning, and, as such, an explanation of its intentions may be expressed as follows:

Recently the R.I.B.A. has paid more and more attention to the subject of co-operation between the building professions, and has sought to increase understanding by each member of the building team of the others' problems, in an effort to create a harmony of interest which should result in more coherent and integrated architecture. At the same time, it is not the aim to treat any allied subject in such detail as to deflect the architect from his true course – the pursuit of harmonious and relevant design – by inflicting on him the intricacies of the complete process of servicing a building. The architect must know enough to be able to consider the requirements of services at the earliest stages and to start out on his design with an in-built understanding of the ways in which his earliest decisions can condition, or be conditioned by, the climatic, thermal and functional environment he sets out to produce.

The shape of this book has been dictated by the pattern of lectures the student may meet during his studies. Clearly, however, individual lecturers will wish to approach the subject in their own way, and the sequence we have adopted may have to be discarded. Nevertheless, we progress from the simple to the complex in a manner designed to give the student as complete an understanding of the subject as possible, according to the level of his architectural studies. Thus, for the third-year student, who may be taking this subject for the first time, we endeavour to give an overall grasp of what environmental-control systems seek to achieve, and the basic information to enable him to carry out simple calculations related to the type of structures he may be designing. For more advanced students, the complex problems of air-conditioning are touched on, but further reading of a deeper nature will be necessary for the student who wishes to investigate that fascinating subject more fully.

Since it has been decided by Parliament that a metric system is to be adopted in the United Kingdom, all facets of the building industry are faced with a problem of change of units. Thus, according to the recommendations of the British Standards Institution, the Système International d'Unités is to be the standard system of measurement. This system has as its basis the following units: metre, kilogram, second and ampere (MKSA). Other units, both basic and derived, exist and can be ascertained by reference to PD5686 issued by the British Standards Institution.

This book uses Système International (S.I.) units throughout, so as to familiarize students with the system, but, to assist reference to other works using obsolescent units, parallel calculations or dimensional approximations in Imperial units are given in brackets. Students should note, however, that where round figures are used in S.I. units they will not necessarily be mathematically correct transpositions of the figures given in Imperial units, since the latter will also be rounded-off. For more exact conversions, reference should be made to the conversion charts (Charts A and B) on pages 210 to 213, in *Appendix* C.

We wish to acknowledge the help and encouragement given by Mr. W. G. Cutting, Senior Lecturer at the School of Architecture, Kingston College of Art, and the assistance received from the Institution of Heating and Ventilating Engineers, as well as all those who kindly gave permission for the use or reproduction of diagrams and tables appearing in this book.

J. N. C.
H. R. C.

London, 1968

Contents

CHAPTER 15: THE FULLY-CONTROLLED ENVIRONMENT 174
Basic considerations. Secondary heat transfer systems. Considerations for selection of appropriate system. Building details to assist system design. Primary generation of cooling: basic principles; room conditioners; centralized systems. Description of components; space allocation for plant. Thermostatic control of systems. Plant location and accessibility. Example of the distribution of plant about a building. Overall design procedure for thermal environmental systems. Gross energy requirement for buildings.

Illustrations

Figure numbers precede captions; page numbers follow them.

1: Human Reactions to Environment

ANY DEFINITION OF ARCHITECTURE would be sure to be disputed by some, but few would argue that any comprehensive definition should not include a reference to the control of the internal conditions in buildings. For we need only look back to primeval men and consider why he started to enclose space to realize that, basically, building is the creation of structures whose internal environment varies from the external climate. Today, buildings are mostly intended for human use, although some are also for animals and plants, and the remainder are for machines. Each use demands certain degrees of controlled conditions or internal environment. This book considers only human use, since this is not only the most common but also the most demanding.

Early man, faced with the severity of the natural elements, first of all clothed himself to keep out the cold or to shield himself from extreme heat. He took shelter in trees, in natural caves or animal burrows from the rain and wind. Later, he adapted the caves and then started to build artificial shelters. Why did he have to do this? Because metabolism (the rate of operation and the capacity of his heat and energy-balancing systems) could only cope within narrow temperature limits, and his efforts were directed at conserving his energy by producing a less harsh internal environment for the times when he was not forced to expose himself to the external climate.

There are therefore two elementary factors which must be considered in providing for human comfort:

External Climate – producing transient heat gain or loss due to the effects of wind, rain, solar gain, etc.

Internal Thermal Environment – relatively fixed conditions of controlled temperature, moisture content of the air and air movements.

Both of these should influence the architect's consideration of the orientation and construction of buildings, and both can vary. The climate of a locality is the synthesis of the day-to-day values of the meteorological elements which affect a locality – e.g. the average and extreme values of precipitation, temperature, humidity, sunshine and wind velocity, and the phenomena resulting such as frosts, cloudiness and soil temperature.

The internal thermal environment results from the control of local (i.e. internal or part of an internal volume) conditions of pressure, temperature, moisture, oxygen and air movement within acceptable limits – the limits being determined by the physiological and psychological requirements of humans – so as to maintain body-heat emission in balance with its surroundings.

Modern man, sophisticated in his desires, will accept even less rigorous conditions. Though his metabolism has probably remained much the same (it can be argued that, with less dramatic physical demands on his system, the rate may have slowed, partly accounting for generally greater longevity), he sees no need to run his system at peak load and, having the energy sources available, now demands the most precise control of his environment.

The control of thermal environment, then, is what we are talking about. The exact means will come later. First, we must look at what the environment consists of, know how to measure what we find and consider in general terms what can be done to modify the constituents, so as to alter the natural state to an artificial one corresponding to the comfort needs of the human beings concerned and related to the functions of the space concerned. Note that we refer to 'the human beings concerned'. This is because different individuals have different levels of toleration of heat or cold, not due merely to their state of health but due also, and more widely, to their metabolic rate. If you have ever held a small bird in your hand and felt the rapid pulse and the warmth of its body, you will realize at what a pace the bodily systems of some creatures function. Their rate of heat conversion per unit of weight is much higher than that of humans, and their food requirements are therefore relatively greater and their capacity to dissipate energy much higher. Birds are among the highest in metabolic rates, the slow-moving and ponderous creatures being among the lowest. Humans fall somewhere in-between, although all are not the same: the 'thin-skinned' individual who is highly susceptible to temperature extremes reacts quite differently to environmental changes than does the hardy outdoor character, and his discomfort is a real physical or psycho-physical reaction. Let us, then, look at the bodily reactions which are related to environmental adaptation.

Subjective reactions of humans to immediate environment

'Feeling cold' is a state whereby the body gives up its heat to the atmosphere more rapidly than the body can produce or distribute replacement energy (from digestive or conversion processes). The bodily reaction to this condition is the contraction of the blood vessels beneath the skin so as to reduce the rate of heat loss, the spasmodic stimulation of muscles (shivering) to increase the rate of heat production, and the erection of the hairs on the skin to trap air and reduce heat loss (goose pimples). Briefly, less heat loss, more heat production.

'Feeling hot' is the reverse condition whereby the heat gained from the surroundings tends to raise the body temperature above its correct level. If protracted, this condition results in heat exhaustion – a form of collapse. The body reacts to the condition by expanding the surface blood vessels to dissipate heat more rapidly, and by the release of perspiration whose evaporation causes a drop in skin temperature. Briefly, more heat loss.

'Stuffiness' is a condition of the atmosphere where there is little air movement and a low oxygen content (air pollution due to smoking, and so on, gives a similar effect). The effect on humans is of lassitude, a slowing-down of reactions due to the low availability of oxygen. This condition depends less on the temperature than on atmospheric conditions, but it is aggravated by warmth.

'Dryness' of the air increases the rate of moisture loss from the body and respiratory system and causes irritation. The skin temperature ultimately tends to rise, due to a drop in perspiration rate. Relief is obviously sought in increasing the intake of fluids to replace losses.

'Dampness' of the air causes discomfort in two ways. In saturated warm air, perspiration fails to evaporate, so the body temperature rises and the skin remains wet. In damp cold air, the moisture absorbs heat from the surface of the body by latent evaporation, thus cooling the body rapidly and producing discomfort. Increased air movement exaggerates this effect.

2: Elementary Heat Transfer

IN DISCUSSING BODILY REACTIONS to varying conditions, we spoke of heat and cold, dampness and dryness in non-technical terms but without defining what these really are. In all scientific studies, the first aim is to isolate the fundamentals of the subject in order to measure and calculate, but, since some of the definitions in physics can become very complex, we have used simpler conventional explanations.

Heat is the energy given off by a material due to its atomic activity. Energy, breaking free from a material, can be *radiated* – much as light is – through the air. Like light, this radiant heat can pass through transparent substances or can be reflected by mirror surfaces. The heat rays are those below red light in frequency. On striking opaque substances with matt surfaces, the energy is absorbed into the material, exciting its structure to a condition similar to that of the material which emitted the heat rays. This 'jostling' of the molecular structure of the material transmits the energy through the material rather like a billiards ball striking another. In this way, the energy is *conducted* through the material. The transmittance of heat energy is a form of work and can be expressed in terms of energy expended over a period of time.

A substance subjected to heating or cooling expands or contracts and its density (weight per unit volume) therefore varies. In fluids and gases, the heating of part of a volume of the substance causes that part to become lighter than the remainder and therefore to rise, setting up thermal currents within the volume. Heating a container of water at its base causes a column of heated water to rise at the centre and to spread out over the surface, thereby losing heat and causing it to drop near the sides, to curl in again and to displace the water at the centre. This is called *convection*. It occurs whenever a fluid or gas is heated in 'static' conditions (such as in an enclosed container or

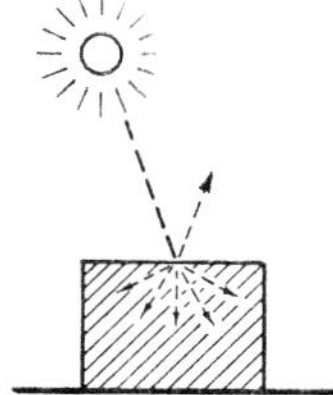

RADIATION : radiant energy, impinging on an object, is partly reflected and partly absorbed, according to the nature of the surface finish. Absorbed heat travels through the material of the object by CONDUCTION, the transmission being like the jostling of molecule against molecule.

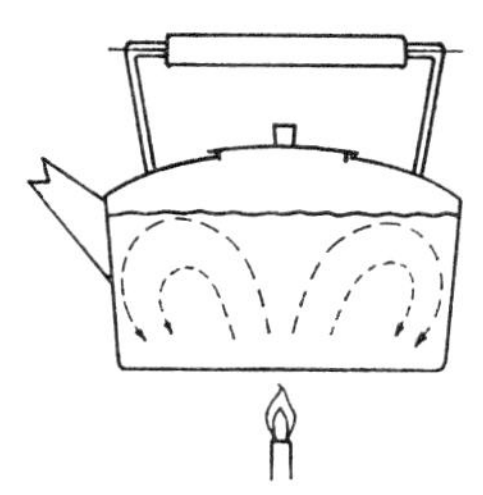

CONVECTION results from the reduction in density of a heated fluid whereby it rises through the cooler, denser fluid. The lighter hot fluid gives up heat by conduction to the cooler fluid as it rises – and to the atmosphere above. It therefore becomes more dense, and drops down near the sides to be reheated... and so on...

FIG. 2.1. Radiation, Conduction and Convection.

room), the column of rising (warmed) material being situated above the heating source.

Thus, we have three elementary methods of transmitting heat:

(i) Radiation
(ii) Conduction
(iii) Convection.

There is also a fourth method: latent heat loss or gain, resulting from change of state, e.g. evaporation.

Cold is a negative term. It represents the absence of heat or the relatively lower temperature of one body as compared with another. As the temperature of a substance decreases, it contracts and becomes more dense as its molecular activity declines. For the purposes of heating, anything below 5° Celsius is cold, while of course anything below 0° is freezing! However, this subjective scale of values results from the fact that water is the most important fluid in our lives and the behaviour of water at various temperatures affects us closely.

In referring to the subjective evaluation of heat, we have touched on the corner-stone of the problem of environmental control – the creation of conditions which appear comfortable to the person using the space concerned. So, let us next consider man in relation to an acceptable thermal environment.

Man is a warm-blooded animal capable of quite high energy outputs for short spells, but given to manipulative occupations rather than to labouring pursuits. (That is to say that the tendency today is the use of skills and artifice, while allowing machines to do the heavy or repetitive work.) *Metabolism* is an expression for the amount of energy a human being can dissipate and, for an average adult in normal temperate conditions, is around 300 watts. This heat loss occurs in accordance with the elementary forms, namely: heat radiated from the body; heat lost by convection from the surface of the body; heat conducted by direct contact, and, finally, heat lost by evaporation of perspiration and the exhalation of air from the lungs. The body makes up for this heat loss by intake of energy-producing foods in order to sustain the body temperature and to store energy against future requirements. It is possible to control the loss from the first four causes, though heat loss from respiration is virtually unalterable.

Clothing is an obvious and a first means of tempering the sharper edges of the climate so as to reduce energy loss and balance the system. In cold weather, when the atmosphere is able to absorb heat rapidly from the body, thus increasing the heat loss, clothing acts as an insulating blanket, preventing the rapid loss and enabling the body to operate at or near its normal level. In hot weather, when the main problem is one of high heat gain from the sun or air, clothing again acts as an insulator, reducing the inflow of heat by radiation and conduction to the body, to obtain a balance at a near-normal rate by radiation and perspiration.

Energy manifests itself in several ways – heat being one of them, work being another. The moving of an object involves work; that is to say that energy is put into the object by the force moving the object. Man eats and absorbs the locked-up energy in the food, converting it by chemical changes into forms which can be used to supply energy to the body tissue. This conversion process itself produces heat to maintain body-temperature, and subsequent muscular activity stimulates the emission of energy and produces heat. Thus, man at rest produces only a small amount of heat, while heavy activity causes an increase in heat output. Moving on from heat-output caused by work to the provision of comfortable conditions in which to carry out such work, it will be clear that the greater the human heat-output due to muscular activity, the less the ambient temperature need be. Thus, the clerk at his desk or the architect on his board needs more heat in his office than the joiner sawing wood. Equally, the joiner needs more food energy for his task than the sedentary worker. (Note, however, that the efficiency of energy conversion of each individual controls the amount of food needed, too; and habit is a factor in determining the quantity *desired* as opposed to *required*.)

Clothing is a considerable factor in determining bodily comfort and in some cases is the only way of maintaining reasonable conditions for industrial processes. Take, for example, the man whose job is sprinkling enamel powder on the cherry-red hot casting in a bath-manufacturing works. The bath is at 800°C, or thereabouts. The enamel must be sprinkled on smoothly by a skilled operative who is subjected to the tremendous heat radiated by the bath. Clothing of asbestos with foil reflectors and insulating padding protects the man, since it is not possible to alter the basic condition of high radiation. Similarly, but at the other end of the scale, cold storage demands personal protection in the form of kapok-filled clothing to retain body heat in sub-zero conditions. In these circumstances, the body is the heat source and loses energy rapidly to its surroundings unless protected.

Where it is possible to extend the 'clothing' or outer protection, so to speak, so as to include a number of persons, a kind of capsule is produced enabling a group to work together irrespective of external conditions. Vehicles are particular examples of this: the saloon car, the aeroplane, the submarine and so on. Buildings, however, are usually required to accommodate a number of varied functions, each of which may require conditions different from the others.

3: Moisture in the Air

LET US RETURN TO THE BASIC FACTORS connected with comfort. So far, we have dealt with heat and cold, but, apart from mentioning the influence of dampness, we have not discussed how the moisture gets into the air, or in what form.

From the spout of a boiling kettle, a jet is emitted; close to the spout is a transparent area; farther out is a visible white mist; still farther out, the mist disappears again. The true steam is the transparent part close to the spout. It is water in a gaseous state. As the temperature (and pressure) drops farther from the spout, the dry steam changes back to very hot water in fine droplets and becomes visible wet 'steam' or vapour. Farther out, still, the water is absorbed into the air and becomes invisible water vapour. Boil the kettle for a long time in a small room and, after a while, the visible steam no longer disappears but hangs about as a mist in the room; at the same time, the windows and other cold surfaces are coated with water droplets. The air has obviously reached a point where it can no longer absorb the water vapour, and condensation has occurred.

Dry air, containing little or no moisture, is like a sponge. It can absorb varying quantities of water vapour according to its temperature. When it reaches a point where it can absorb no more, it is said to be saturated, but the actual quantity of water per unit volume of air is dependent on the temperature – the higher the temperature the greater the capacity to hold water vapour. The quantity of water contained per unit quantity of air is known as the *specific humidity*. Because the actual quantity of water at saturation varies with temperature, the quantity of water in a given volume of air is, for convenience, measured as a percentage of the quantity contained in the same air at saturation, and is known as the *relative humidity*. Thus, while at summer temperatures air containing a given quantity of water may be described as being at, say, 50 per cent R.H. (relative humidity), the same air with the same actual quantity of water per unit volume, but at winter temperatures, may be described as being at 70 per cent R.H. This is because the quantity of water at saturation is lower at lower temperatures, and the quantity of water actually contained represents a higher proportion of that lower quantity.

To recapitulate with the kettle. First, we have saturated steam at the kettle spout. Next, we have visible vapour being absorbed by the air. Then, later still, the vapour is fully absorbed and the humidity of the air has risen. In the case of the vapour-filled room, saturation of the air was reached and the surplus vapour could only hang about as a mist until either the room temperature was raised (to raise the saturation level), or the air was changed to bring in a new quantity of unsaturated air capable of absorbing the surplus vapour. The windows, being of low thermal resistance and exposed to external lower temperatures, were cold and the air in immediate contact with them was colder than the surrounding air (thus reducing the saturation level) and was causing a heavy fall-out of vapour contained in the air. The vapour condenses as visible drops on the cold glass because the surface and the surrounding air are below dewpoint. *Dewpoint* is the point on the thermometer at which air at a given humidity (not relative humidity, but the actual quantity of water per unit volume of air) becomes saturated. Consider a volume of moist air at a high temperature and low relative humidity. Cool the air down and the relative humidity rises. Cool it further and eventually the air will be so cool that it can only just hold the vapour contained in it – i.e. it has reached saturation. This is the dewpoint for that specific humidity (not relative humidity), and any further cooling results in visible water droplets condensing-out in the air and coagulating on cold surfaces. (This, in simplified form, explains moisture content, but, in actuality, humidity is a complex matter involving differing vapour pressures. However, the foregoing covers the subject adequately for ordinary heating purposes.)

Clouds are formed in this way. Warm, moist air rises from the ground and is cooled as it rises. Eventually, it reaches dewpoint and the vapour condenses-out as cloud.

Similarly, at ground-level on a clear night, the heat loss by radiation can be so rapid that the ground cools off before the air and heavy condensation then occurs at ground-level, fol-

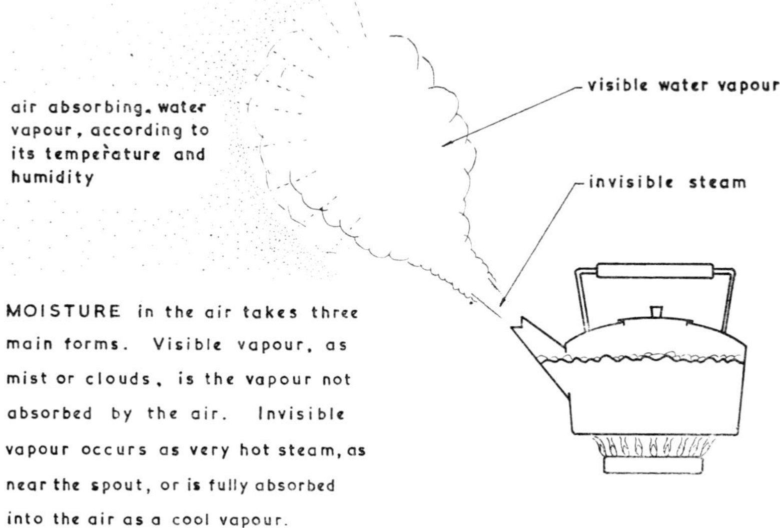

FIG. 3.1. Moisture in the Air.

lowed by a mist rising as the air is cooled below dewpoint by the ground. Condensation also occurs on the cooling coils of a refrigerator, where moisture released from stored goods condenses on the coils, freezes, and must be melted away by 'defrosting'.

If a volume of saturated air is introduced to another volume of dry air, a process of interchange takes place whereby absorbed moisture flows from the wet air to the dry until an even distribution is achieved. This effect is due to the 'vapour pressures' which even out by flow from high pressures to low pressures.

4: Thermal Measurement and Heat Loss

THE ORDINARY THERMOMETER MEASURES, by the measurement of the expansion of mercury or alcohol, the ambient air temperature. This is called the *dry-bulb temperature* and, for external readings, is always taken in the shade. This is because radiant solar heat passing through the glass and striking the fluid tends to raise the temperature above the actual air temperature. In a similar way, the human being is also warmed by radiant heat, even though the air temperature may be lower.

So we have another kind of temperature to consider for comfort, namely *radiant temperature*; and to measure this we use a *globe thermometer*, which is an ordinary bulb thermometer, but having a sphere of blackened copper around the bulb to absorb the radiant heat while preventing the air from reaching the bulb. This thermometer records the globe thermometer temperature and, since radiant heat varies with the type of emitting surface (e.g. the surface temperature of glass is continually changing, whereas thick masonry wall-surfaces change only slowly), the readings, when associated with air movement and dry bulb temperature, produce the *mean radiant temperature.*

Having discussed humidities in the last chapter, we can now consider how to measure them. Clearly, one could cool down a known volume of air, condense-out the water vapour and find out how much there was in that volume and so calculate the humidity. This is not very convenient or practical, however, and what is done in practice is to measure the rate of evaporation of water by its cooling effect on a thermometer and compare the reading of this cooled thermometer with that of an ordinary thermometer.

A wick, with one end in a container of distilled water, is wrapped around the bulb of a thermometer and exposed to the air-movement (or swung about) to set up evaporation of the water, which cools the bulb and gives a lower reading than the similar, but dry, bulb thermometer. The difference (known as *wet-bulb depression*) between the two readings forms the basis of the calculation of the relative humidity. When the wet-bulb reading equals the dry-bulb reading, the air is at saturation since no evaporation is taking place, indicating that the air can absorb no more vapour.

Temperature is the level of activity of heat energy, but it is not a measure of the quantity of heat. To measure heat, a unit is needed describing the actual quantity of energy. As seemed best when creating units, the scientists took a unit volume of a common substance (water), found out the amount of heat to raise it through a unit of the temperature scale, and called this the *unit of heat*. Currently, in the U.K., we use a pound of water and one degree Fahrenheit, whilst the Continent uses a gram of water (1 c.c.) and one degree Centigrade. So we have two quite different units – the *British Thermal Unit* (Btu) and the *Calorie*.

However, in the light of the Government decision to adopt a metric system, the I.H.V.E. committee appointed to deal with the problems of the change have recommended the use of the Système International d'Unités (S.I.) – a system which is gaining acceptance on the Continent. In this system, the unit of heat (or energy) quantity is the *joule* (as used in electrical calculation). The joule being a small quantity, it is recommended that the *kilojoule* (kJ), representing 1000 joules, will be more convenient in practice.

$$1 \text{ kJ} = 0 \cdot 95 \text{ Btu}$$

Here, therefore, we adopt the S.I. units but, in order to facilitate reference to earlier published information, equivalent values or parallel calculations in obsolete units are also given.

Taking our kettle holding say 1 litre (1000 c.c.) of water at 5° Celsius (in general terms interchangeable with Centigrade), how much heat do we need to put in to heat it to 15°C?

The simple expression for a change of heat content is:

Mass (kg) × temperature difference (°C) × specific heat (Cp)*

Thus the heat required in our case is:

$$1 \text{ kg} \times 10°C \times 4180 \text{ (specific heat)} = 41,800 \text{ joules}$$
$$= 41 \cdot 8 \text{ kJ}$$

Or, in the current U.K. terms:

$$2 \cdot 205 \text{ lb} \times 18°F \times 1 \text{ (specific heat)} = 39 \cdot 69 \text{ Btu}$$

* See the Equivalents Table, pages 210 and 211.

These quantities take no account of the time taken for the change to occur. Time is an essential part of heat calculations for buildings and, when incorporated in the equation, represents the thermal dynamics of the building, since these are variable with the time of day or year. The rate of thermal gain or loss is therefore an expression of the energy absorbed or emitted by a building and its occupants.

For example, if it took one minute to raise the water in the kettle to the higher temperature, the rate of energy gain would be represented as:

$$41 \cdot 8 \text{ kJ per 60 seconds} = 0 \cdot 6966 \text{ kJ/sec}$$
$$= 0 \cdot 696 \text{ kilowatts}$$

Or, in current U.K. terms:

$$39 \cdot 69 \text{ Btu per } \tfrac{1}{60} \text{ hour} = 2381 \cdot 4 \text{ Btu/hr}$$

The rate of heat flow depends upon the resistance a material offers to the flow, and this is closely related to the density of the material. Most metals conduct heat rapidly, while discontinuous materials such as foam rubber or expanded plastics resist the flow of heat, and, since the bulk of the heat loss from buildings results from conducted heat flow, it is clear that the rate of loss must be ascertained.

For each material there is a value of conductivity which can be established by experiment. This value, given the symbol 'k', corresponds to the number of joules (Btu) which will pass through one square metre (one square foot) of one metre (one inch) thickness of the material in one second (one hour) when there is a temperature difference of one degree Celsius (one degree Fahrenheit) between the two surfaces. Or, put into the form of equations:

$$\text{'k'} = \frac{\text{J m}}{\text{m}^2 \text{ s deg C}} \qquad \left(\frac{\text{Btu in}}{\text{ft}^2 \text{ hr deg F}} \right)$$

And, since 1 joule per second equals 1 watt,

$$\text{'k'} = \frac{\text{W m}}{\text{m}^2 \text{ deg C}}$$

Most building materials have been tested by the National Physical Laboratory, or other recognized body, to establish their values of 'k', and, for materials used in walling and roofing etc., the value is given as one of the characteristics of the material. It is all that is needed to ascertain, for a given thickness and temperature difference, the amount of heat which will be conducted through a material. For example, supposing a material has a 'k' of 0·007 Wm/m² deg C and a thickness of 0·05 m. Its conductance would be $\dfrac{0·007}{0·05} = 0·14$ W/m² deg C. If the thickness were doubled, the conductance would be $\dfrac{0·007}{0·10} = 0·07$ W/m² deg C, showing that the thicker the material, the less the conductance.

Thus, from such information it is possible to calculate the heat loss which will occur from a building, by taking each of the materials in, say, a wall and combining their values and multiplying by the area of that type of construction to arrive at an aggregate conductance. Combining the values of each constituent material is not, however, the simple addition of 'k' values. The reciprocal of conductivity is resistivity ('r'), and we must add the resistivities and then find the reciprocal of the total to establish the total loss for a complex construction. To allow for surface exposure, we must add in a resistance figure for each of the inner and outer skins, since there is a 'skin' resistance to heat flow at these surfaces. Thus, our calculation can now be expressed:

Total heat transfer per unit area of construction in unit time:

$$\frac{1}{\text{Rsi} + r_1 + r_2 + r_3 \text{ etc.} + \text{Rso}} \text{ W/m}^2 \text{ deg C}$$

The value is represented by the symbol U.

It will be seen that resistivity is more useful than conductivity in these calculations, and it is therefore proposed to adopt 'r', in S.I. units, in calculating overall heat transfer.

To find the total heat loss of a structure, we multiply the area of surface by the U value and by the temperature difference between the interior of the structure and the exterior. Now, here is where a choice can come into the problem. Usually, the external temperature is taken as $-1°$C (30°F), but the interior temperature is a matter of selection or of statutory requirement, often taken as 18° (65°F), thus giving a difference of 19°C (35°F).

Since heat loss is measured in J/sec – that is, watts (Btu/hr) – it is logical that heat input should also be measured in the same terms; so it can be seen that from the heat loss calculation it is possible to establish a boiler size, and, provided the actual heating equipment is rated in a similar fashion, it is simple enough to ascertain the size of, say, a radiator to suit the heat loss of a given location.

Another factor affecting heat loss through structures is the exposure of the building. A building sited on an exposed coastline is subject to higher winds than another nestling in a valley, and a building buried in the heart of a large city is protected still further from winds. The effect of exposure to wind is to speed up the rate of heat loss and, therefore, to affect the U value of the wall by reducing the external surface resistance. One can regard it from the viewpoint that the greater quantity of air passing the wall at a normal temperature is equivalent to the normal quantity of air, but at a lower temperature, thus producing effectively a lower external temperature. Conversely, a hot wind in a tropical climate transfers more heat to a building than hot, still air.

Thermal gradients

In considering the passage of heat through a material, we took an internal temperature higher than the external temperature, and set out to show that, according to the materials used, the heat meets greater or lesser resistance to flow. If the temper-

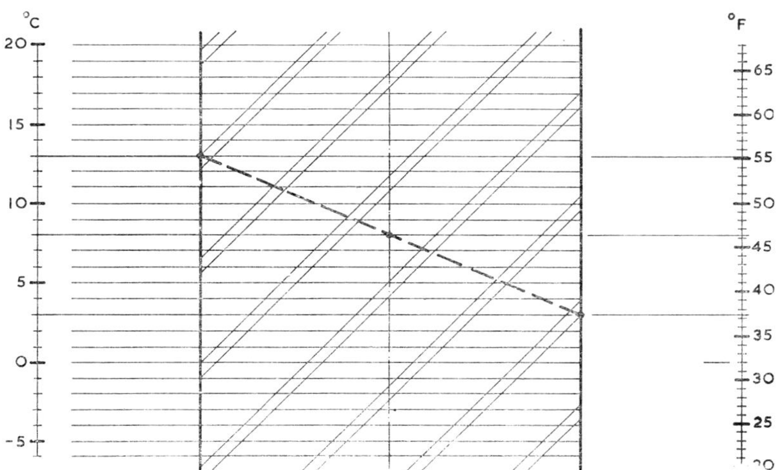

FIG. 4.1. Thermal Gradient through Homogeneous Wall.

ature inside is, say, 15°C (60°F) and the temperature outside is, say, -1°C (30°F), then somewhere in the middle of the material the temperature is the mean of the two. So, if we take a wall whose inner surface is at 13°C (55°F) and outer surface 3°C (37°F), the temperature exactly half-way through the wall will be 8°C (46°F). This is only true if the material of the wall is constant throughout. Thus, it is logical that the thermal gradient of a homogeneous wall will be as illustrated graphically in Fig. 4.1 (page 27).

The effect of surface resistance is illustrated by the kinks in the graph at wall surfaces, while the dot-lines indicate the associated air temperatures, the internal temperature tending to dip near the wall due to the loss of heat to the wall causing a downward convection current cooler than the general temperature in the room.

Few walls or roofs are, however, as simple as this, and the graph is usually a complex mixture of varying gradients whose slopes depend upon the insulating qualities of each material used. Consequently, the thermal gradient for a non-homogeneous wall might appear as in Fig. 4.2.

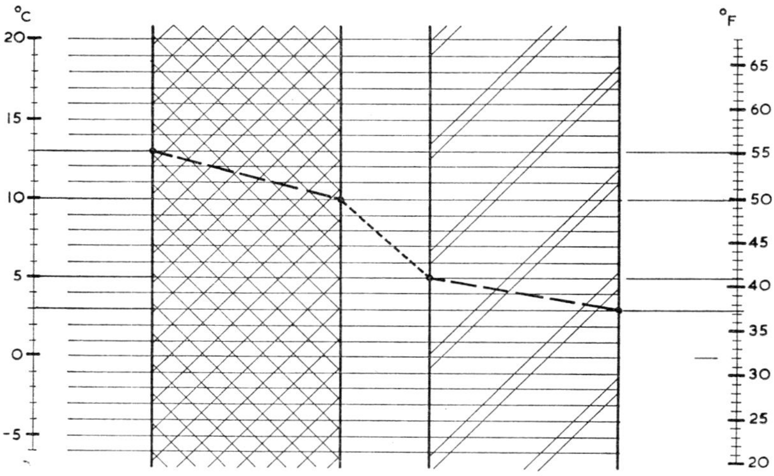

FIG. 4.2.   Thermal Gradient through Cavity Wall.

Note that the inner and outer surface temperatures are the same as for Fig. 4.1; it is the line between which differs and, in practice, indicates the varying insulating quality of the wall.

Let us reconsider what was said earlier: that resistance to heat flow is closely related to density – that is, that low density materials offer a greater resistance than high density materials. Materials with a discontinuous structure, such as plastic foams, are low in structural strengths and are unable to hold much heat because of low mass. Materials of dense structure have high strength and can absorb quantities of heat, though they transmit it readily too. Foam concrete (though strong when compared with expanded polystyrene), is not as strong as brick, but it has a discontinuous structure and offers greater resistance to heat flow. It is therefore a better insulator than brick. This property can be looked at in two ways. Firstly, given a desired internal temperature and an assumed external temperature, the use of a good insulator results in less heat being needed to achieve the internal temperature than would be needed if a poor insulator were used, because the rate of heat flow is slower. Alternatively, given a fixed heat input, a thinner wall of good insulation will achieve the same thermal result as a thicker wall of poor insulation. The choice of materials is not always dictated by thermal considerations, however, as durability, structural strength and appearance must also be considered, and the cost of insulation must be balanced against the capital and running costs of heating.

Radiant heat loss

So far, we have considered only conducted heat and the insulation designed to reduce conduction. In addition, however, all surfaces radiate heat and the higher the temperature the greater the rate of radiation. Radiation from a heating element is absorbed by the room surfaces, conducted through the structure, warming it as it passes and therefore increasing its rate of radiation. As stated earlier, radiant heat behaves like light and can be reflected; thus, to reduce radiant heat loss through walls a polished metallic reflector is used, usually in the form of a foil bonded on to a board or panel. The heat radiated from a material is reflected by the foil back into the emitting material, though heat can still be conducted *via* the foil at points of contact. Bright surfaces, such as foils, also possess low emissivity, so that when fixed to an emitting surface their rate of emission is very low, thus preventing heat flow.

Convected heat loss

Convected heat is not usually a prime cause of heat loss from buildings as most buildings can be totally enclosed to prevent convected heat escaping. But, to take the case of a glass works where the problem is more how to lose heat from the atmosphere than to retain it, the convected heat escapes through large roof ventilators and cooler air is drawn in at low level to replace it. Similarly, the cooling tower of a power station relies largely upon the convection current rising within the tower to cool the hot water from the generating plant.

Convection can, however, assist other effects, such as infiltration, so as to aggravate their influence on thermal environment.

Convection losses can be significant in buildings when the construction uses broad cavities for insulation. In a narrow cavity, though one surface is warm, there is not enough space for any significant convection current to develop; whereas, if the cavity becomes broad, it is possible for air to start circulating within the space so that air on the warm side rises, curls over the top and descends on the cold side, releasing its heat to the cold material and thus increasing the rate of heat loss of the construction.

Heat loss to the ground

So far we have dealt with those surfaces of buildings which are exposed to the external air, but, remembering the principle that heat flows from hot to cold, it will be realized that part of a building structure in contact with the ground may also lose heat by conduction into the ground itself. In this case, the exposure of the building is not a matter of hills and valleys but is more a question of the nature of the soil and its capacity to absorb heat.

Rock, being a dense and highly integrated material, can conduct heat rapidly, whereas a gravel soil, due to the air spaces between particles, will not. On the other hand, dry sand is a fair insulator while wet sand is not. (The water, in effect, gives continuity to the otherwise discontinuous structure.) Thus, the various qualities of the type of soil concerned will give rise to greater or lesser heat losses from the building, and we have thereby come up against the last major factor in the subject of heat gain and heat loss – thermal capacity.

Thermal capacity

To find the obsolete units of heat, a fixed quantity of water was raised by a fixed temperature increment. But suppose that, instead of water, some other substance had been chosen: lead for example, or oxygen (one could not choose concrete or any other variable material). The result might still have been a calorie or a Btu, but the actual quantity of heat energy represented by the unit would have been different – for it is fairly obvious that, if a heavy dense material like lead had been chosen, the quantity of heat energy would have been larger, and that if a gas had been chosen the quantity would have been smaller.

So, we arrive at the conclusion that to produce a given thermal change different materials require different quantities of heat energy, and, combining this with our earlier discussions on heat flow, it is possible to produce a series of types of material whose thermal capacities follow a pattern as follows:

(*a*) Dense, continuous materials (e.g. metals, hard rocks, dense concrete): high thermal capacity; rapid conduction.

(*b*) Fairly dense but discontinuous materials (e.g. foam concrete, wood, gravel, soils): medium thermal capacity; slow conduction.

(*c*) Low density, discontinuous materials (e.g. insulation board, foam plastics): low thermal capacity; very slow conduction.

Note that this crude breakdown is not intended to be flawless, but it is sufficiently true to be applicable to most building problems, as is the basic premise that the higher the density the greater the thermal capacity.

Thus, a building constructed of dense and heavy materials requires more heat to raise the fabric temperature than a lightweight building, but the heavy building will store that heat and release it after the applied heating ceases, whereas the light building has no stored heat and therefore cools off rapidly.

This lag of temperature-response due to the use of materials of high thermal capacity is sometimes known as the 'thermal inertia' or 'thermal time-lag' of a building, and it can play an important part in heating calculations. (See *Post-War Building Study*, No. 33.)

We are now at a point where it is possible to study the internal environmental-requirements of a building in relation to its external climate and consider the selection of the type of enclosing fabric which will best serve its needs. The architect, however, does not make his choice on heating grounds alone. Leaving aside the philosophical aspects of design and expression, he also takes into account the structural and functional demands of the building, applying the appropriate weight to them so as to arrive at a suitable and economic blending of performance-characteristics most suited to that particular project.

But, before proceeding further, it is necessary to study the characteristics of the external climate, which is the cause in the first place of the need to control environment.

5: Climate

IN THE PREVIOUS CHAPTERS we have tended to look at man as a creature in his environment, analysing some of his needs and the means he uses to measure them, and studying the properties of the materials he uses to satisfy them. The study of architecture is a sociological study and it is therefore rightly concerned with the things man does and feels, but the practice of architecture includes the fulfilment of man's needs in the context of the prevailing conditions surrounding a building. Let us, therefore, consider the subject of climate and try to establish what are the factors which create the need to control the internal atmosphere of buildings.

The Earth is a rotating spheroid. The rotation is from west to east and the atmosphere, a thin shell of air and water vapour, goes round with it.

At the Tropics, the sun is overhead for one instant of time each year; between the Tropics, the sun passes overhead twice each year. Thus the tropical belt is the area of surface which receives the most concentrated solar heating, while outside it lie the north and south zones where, due to the curvature of the Earth, the sun strikes with slanting rays, thus spreading its heat more thinly. As the latitude increases, the angle of the sun's zenith becomes less and less until a latitude is reached where, at one time of the year, the sun does not rise above the horizon. These latitudes are the polar circles.

We have seen that heated air rises. In the tropical zone, the solar heating causes a great upsurge of air which spreads out as it rises and cools as it travels north and south to descend thousands of miles away. At the same time, air moves in to replace the rising air. The complicated sequences of air movement, allied with the eddies and secondary currents of the movement of heavy (cold) polar air, create an overall pattern of wind and weather.

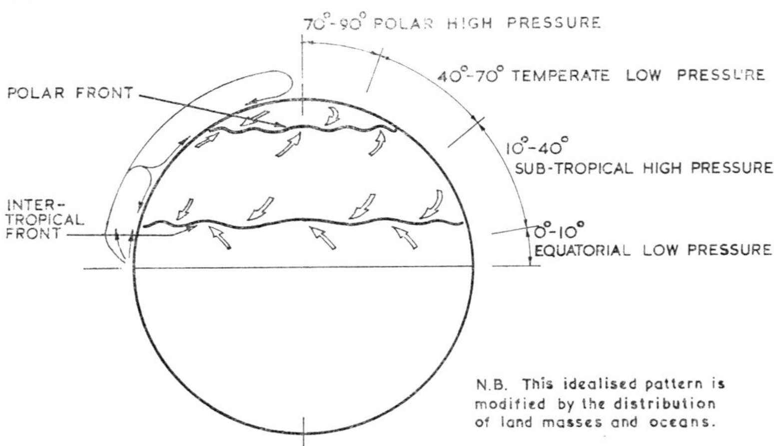

FIG. 5.1.   General Circulation of Winds (diagrammatic).

These air movements take place within the dozen kilometres
or so of thickness of the troposphere, but are on a tremendous
scale of distance. Fig. 5.1, the vertical scale of which is greatly
exaggerated, illustrates the general pattern, and also shows
another important factor – the existence of two areas (the
Fronts), where high-pressure airs meet low-pressure airs. As can
be readily appreciated, the movement of the sun's path up and
down the tropical zone as the year progresses moves the point
from which the major upcurrent springs, and this movement
affects the latitude of these 'Fronts' seasonally, too.

Comparison of Figs. 5.2 and 5.3 shows this seasonal shift of
the equatorial rising current. The whole system, operating on a
sphere of uniform surface, would set up a regular pattern of air
currents and bands of virtually unchanging climate. But the
surface configuration of the Earth is by no means uniform: the
greater part is water, while the land masses are irregular and
tend to be more heavily concentrated in the northern hemisphere than in the southern. Radiant heat from the sun heats up
the land masses very rapidly, giving rise to strong convection
currents, while the water is slower to heat. Desert areas release
the heat almost as fast as it is absorbed, while forest areas
release their heat more slowly.

All these variations in behaviour give rise to a complex pat-

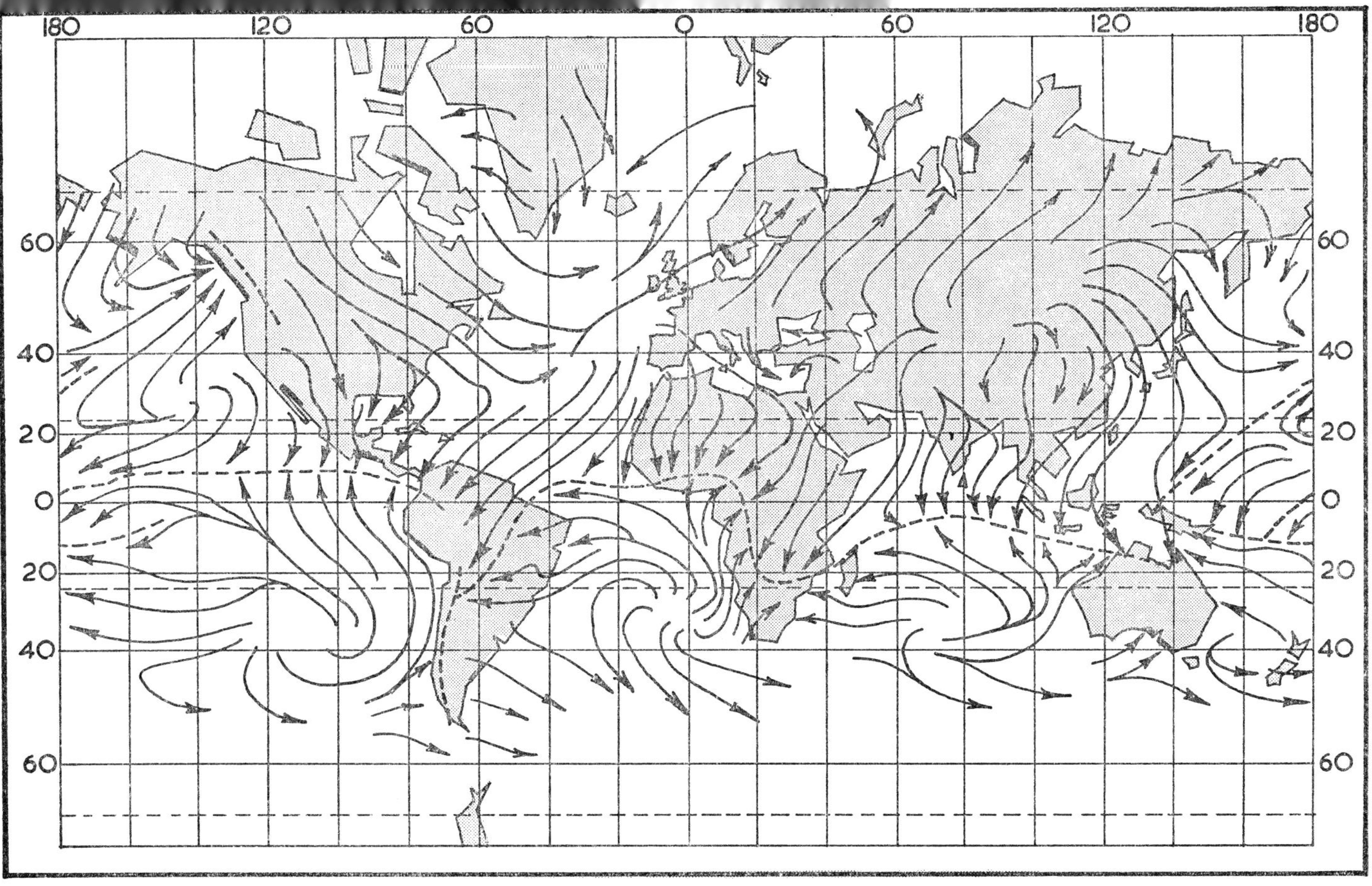

Fig. 5.2. Mean Surface Winds in January.

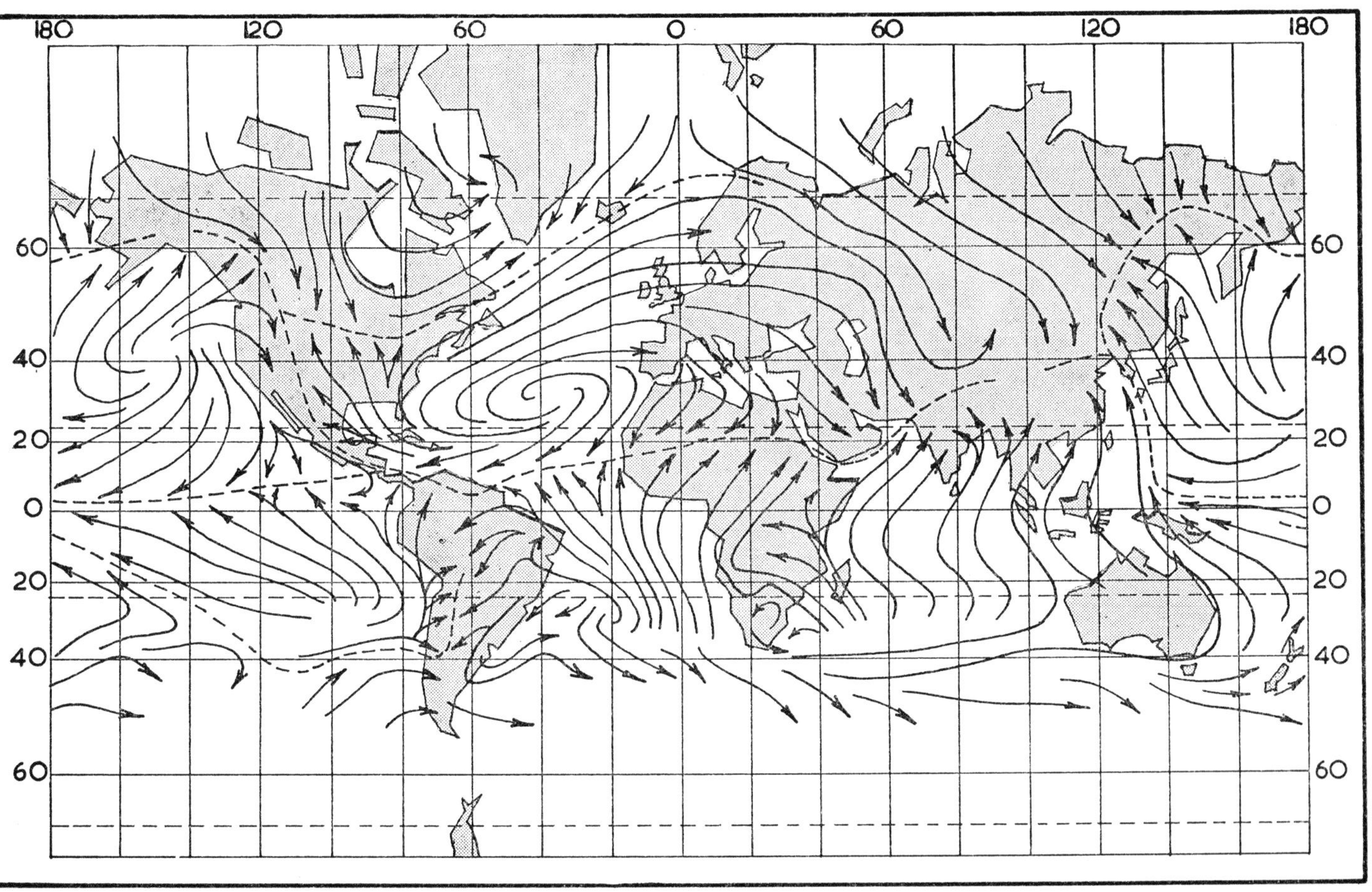

Fig. 5.3.   Mean Surface Winds in July.

tern of air movements, which is still further modified by the density of the airs and their humidity. Warm air, passing over a large stretch of water, absorbs vapour and becomes humid. Warm air passing over a large land mass tends to lose moisture. Cold air cannot hold as much water vapour as warm air but is more dense and therefore heavier. It is the combination of these types of air, meeting each other as they swirl around, driven by the basic global pattern, which sets up the sequence of low and high pressures which drift in on us from the Atlantic.

The British Isles

The British Isles lie off the west edge of a large mass and on the east margin of an ocean. They lie just below the Arctic Circle on the northern fringes of the unstable zone. Taking the west-to-east tendency of weather, it is clear that most of our air comes from the Atlantic and tends to be humid. In winter, when the sun is near the southern Tropic, a belt of depressions passes over the country, while in summer this belt moves northward with the global convection system.

The presence of the Gulf Stream, or at least the drift of water generated by it, ensures that, despite our fairly high latitude, the airs reaching us are not excessively cold; but, when a large high pressure system (anticyclone) lies over the area, the cold air from Europe and Northern Asia drops temperatures for long spells. However, the clear skies associated with a high barometer allow the sun to warm the surface, though the air stays cold.

The physical geography of the British Isles affects the pattern of weather. Western Ireland, rocky and hilly, takes the worst of the Atlantic rain while Cornwall and the Hebrides enjoy a similar plenitude of water. However, a great deal still remains to fall on Wales and Lancashire, though the Pennines ensure that most of the rain comes on the western side, leaving Yorkshire with markedly less. Certain areas to the lee side, notably the south shore of Moray Firth, benefit from the föhn effect whereby a warm dry wind flows down from the Highlands when moist south-westerlies are blowing.

On the whole, the climate is a fairly wet one, reasonably mild and closely like that of British Columbia whose situation in relation to the North American land mass is similar to that of Britain to Europe. There are of course the oddities, like the Hertfordshire frost hollow and the strange mildness of parts of the North Wales coast, but these are specialities of local geography and would need their own explanation.

Wind; fog; exposed localities

Wind, as a term, describes the speed of travel of air, and it is logical that in the open spaces of the ocean where the progress of the air is not impeded speeds are higher than inland. This is only true when we are dealing with wind at surface-level, of course, but the effect of rough terrain extends over considerable areas on the lee side and can significantly alter local conditions. Above such impedances the wind speeds rapidly reach the levels associated with open country, though the effects of turbulence, due to friction and rough terrain, become negligible above 500 metres (approximately 1500 feet).

Apart from the winds associated with weather systems a special case arises at the coast. As we have said earlier the land heats up quickly during the day, sets up convection currents and draws cool air in from the sea. This, subject to the overriding strength of a pressure pattern wind, creates the on-shore day wind noticeable at the coast. At night the position is reversed and the air over the quickly cooled land is drawn off shore by the air currents rising from the relatively warmer sea. Hence, on a calm day, an evening tide is a better sailing time than a morning tide. The strength of on-shore winds is dependent upon the strength of solar heating and the nature of the land surface.

As would be expected, humidities are fairly high in Britain. Mostly they hover around the sixty to seventy per cent R.H. mark, but at certain times of the year, notably in autumn with falling temperatures over land and the Atlantic still relatively warm, the R.H. rises to saturation and mist and fog form. Radiation fog has already been described, but the sort of fog which blows over the notorious Newfoundland Banks, or the fog which blankets large areas of Britain for days arises from *advection*. Advection is the passage of warm moist air over cold ground whereby the temperature of the air near the ground decreases, reducing its dewpoint and resulting in condensation of visible water droplets. The process continues; the layer of fog getting thicker and thicker, until solar heating (much reduced by the fog) warms up the ground and thus the adjacent air. The fog then 'lifts' (i.e. the water droplets evaporate), the process continuing upwards and becoming more rapid as the fog thins. Fog imposes higher thermal loads on buildings due to the greater thermal capacity of water vapour but this is not a dramatic increase, even though fog-water droplets can be 'supercooled' (that is, dropped below freezing point without actually becoming solid) whereby on impact with an object they freeze immediately into ice crystals. The main nuisance value of fog as regards building is the load it imposes on ventilation plant (particularly filters), especially when smoke particles are present.

The effect of terrain on wind speeds has been referred to. Clearly, one could assume that a rocky spur at Land's End would tend to get generally higher wind speeds than a tree-clad hollow in Warwickshire.

In effect the structural engineering profession have already had to face this problem in their design work, and they have, in *Code of Practice*, Chapter V, outlined geographical zones of exposure as well as drawn up a series of conditions graded to establish the exposure of a given site. In this way a reasonable basis for the estimation of the severity of wind conditions can be found. Similarly, a guide to the local conditions to be anticipated in various localities is given in the current *I.H.V.E. Guide*.

But there may be local conditions which have a powerful influence over a small area and which require more than generalized formulae to foresee. For instance, as has been outlined, most of our wind comes from the south-west and most of the strongest winds come from this quarter, therefore a valley which lies on an axis parallel with the prevailing wind will tend to allow the wind to pass through virtually unhindered, while a valley on an axis across the line of air flow will tend to have lower-speed winds but greater turbulence due to eddies curling over the crests of the enclosing hills. To pursue the argument further still, if the south-westerly valley is wide at the south-west end and narrow at the north-east end, wind speeds will rise as the air is constricted in the narrow space. Such wind funnels as this are not uncommon and, if the valley terminates as a blind alley, the enclosing hills can be not only windy but subject to very heavy rain, too, as the air is forced up sharply to escape.

The turbulence arising from undulating ground is less important from the exposure point of view than it is as regards smoke dispersal and flue conditions. The wind current that blows down the slope tends to be variable and gusty. Such down-draughts can cause fires to burn badly with smoke blowing back into the room, or boilers to burn with low efficiency and heavy fumes. Buildings sited in vulnerable places have tall flues with cowls to protect the fire from the down-draught. An up-draught, on the other hand, tends to induce draughts of higher speed in flues, causing problems in the furnace.

Another effect of the conformation of the ground is more abstruse than that outlined above and results from the subtle pressures exerted by hills and valleys. Perhaps a brief word here on the nature of storms is not out of place. A storm is a pressure disturbance in which strong vertical currents of air move violently up and down within the tall cloud formation. The whole system is a poised balance which can be powerful enough to damage buildings, but is nevertheless subject to deflection by variation in contour. With the fairly regular westerly and south-westerly origin of our rainstorms, this effect leads to the formation of storm tracks whereby certain areas frequently receive storms while nearby areas miss them. Only recently have such 'tracks' begun to be defined by meteorologists, and local observation is usually the only source of information.

Location of towns

Because of these various factors, naturally growing towns (as opposed to artificially planted communities) tend to have developed in sheltered locations when other factors such as water supplies or defensive requirements have not overridden. The layout of such towns developing over the centuries is based partly on the prevailing weather, and streets tend to be run so as to break the force of the wind. Such is not necessarily the case today when town expansion takes place largely on a basis of expediency and the feasibility or economy of extending existing services. New towns tend to be sited regionally for political-sociological reasons, and locally for road pattern and land fertility reasons. Exposure and local patterns of weather are minor factors and have therefore usually to be overcome by suitable design of the buildings themselves. Thus, though in some cases the architect can avail himself of favourable factors, he is usually forced to take the site given him and do his best to ensure the correct performance of his building despite the elements; but to overlook these factors in the design of a building can result in high running costs and low functional efficiency in the normal use of such a building.

6: Towns, Sites, Buildings

WE HAVE NOW STUDIED SOME OF THE principal weather factors which finally give the site of a town its character as far as
external influences are concerned. Let us then consider some
other factors related to towns. There is a fairly regular pattern
of the larger settlements which can be discerned and which
evolved principally during the last century when, as the indus-
trialization of the country proceeded, steam power and chemical
industry developed. Because of the prevailing south-westerlies
the smoke from factory boilers and railways and the smells of
the dyeworks and paint-factories drifted north-east, and, therefore, though the poorer people had to live near their work in the
eastern and north-eastern sectors, the tendency of towns was to
spread south and south-west with dormitory suburbs, so as to
be to the windward of smoke and smell sources. Nevertheless,
local geography causes many deviations from this pattern.
Coast towns tend, during the day, to benefit from the on-shore
breeze which affects the westerly trend on eastern coasts. New-
castle, for instance, has its expensive residential sector north-
east of the city centre, high up and mainly clear of the chan-
nelled winds of the gorge-like Tyne Valley.

As far as control of the internal climate is concerned, these
town patterns have a limited influence, but it will be clear that a
location on the leeward side of a town may enjoy less strong
winds due to the breaking up of the wind force by buildings and,
in fact, the centre of a large conurbation may be so sheltered
that its climate differs from that of the suburbs. The centre of
London benefits in this way, with average temperatures as much
as 2°C above suburban temperatures. At the same time, towns
are not as polluted as in the past. Control of smoke emission
and chimney-heights and the pressure towards more refined
fuels is steadily improving the atmosphere of cities, so that to be
on the leeward side does not entail the penalties it used to do.

Similarly, the diversification of industry and its growth in light manufacturing trades have necessitated the embedding of small industrial areas within the generally residential districts, and the enlargement of towns has swallowed up outlying settlements lock, stock and barrel, so that the south-western fringe of one centre merges into the north-east margin of another.

Site characteristics

Within this variegated pattern, the architect may find his site and have to analyse the various weather aspects of it in order to get the best out of it. For several reasons it is useful to study the surroundings of the site, noting the ditch layout, the wind-break pattern of farms, the lean of the trees and hedges and so on. It is also useful to enquire locally into such matters as flood levels and soil conditions so as to establish a basis from which the logic of the climatic-control system may develop.

Bearing in mind the overall weather trends, trees on a western boundary may prove a valuable asset, or perhaps the presence of land-drainage a timely warning. The built-up site will not often be as revealing as the more rural site, but the surrounding buildings may tell a story of driving rain or damp soil, or may provide some shelter of which advantage may be taken in both planning the building and organizing its servicing. But in an age of obsolescence one cannot rely upon the continued presence of shelter derived from adjoining sites and it may be necessary to disregard such factors.

A recent study of the sheltering effects of various barriers clearly showed that the permeable screening afforded by a contiguous tree belt gave shelter (i.e. reduced wind speeds) for greater leeward distances than the solid barrier created by building structures. The resistance offered by the open mesh of branches and leaves slowed down the speed but did not cause dramatic pressure changes, where the solid barrier gave negligible winds immediately to leeward but gave rise to very high-speed currents over the top edge, with heavy turbulence behind. The recovery to original wind speed at ground-level occurred at a distance to the lee of the barrier equivalent to only about twice the height of the barrier.

Thus, the effect of different kinds of screening should be gauged, bearing in mind that most of our rain (causing a reduction in insulation value of porous structures) comes from the south-west and west; our coldest, but often drier, winds come from east and south-east; most of our snow comes from the north, north-west and west.

Tall buildings

The height of a building has already been included as a factor in determining design wind-speed. This, added to the reduction in shelter at high levels, has a considerable influence on the heat loss from buildings, and it may be considered, for example, that a non-porous western wall will be essential if thermal insulation is to remain at a reasonably constant level when the height of a building reduces shelter and exposes the wall to the worst and wettest that we get. But the porosity of a wall creates another characteristic, too. It is difficult to conceive of a material like brick being subject to the passage of air but, in fact, all porous materials (but not necessarily closed-cell materials) allow the passage of air in quantities depending on the openness of structure and the pressure differential on each side of the membranes. This effect is known as *infiltration*. Added to the air passing through the material, there is, of course, the air which blows in through cracks, at doors and windows, through air bricks and between roofing tiles, plus the draught induced by the 'stack' effect of tall buildings. The sum total of all this casual air-change can, at high levels, cause a very considerable heating or cooling load in the buildings.

The air-change resulting from infiltration is normally nothing like as great as that from natural ventilation. Though the free-air cross-section of the clearances around a door is about 0·02 m² (30 sq. in), the smallest window opens up more than that area, and with less resistance, too; so the opening of windows can account, at ground-level, for perhaps six air changes per hour. (The *air-change* is a common unit of ventilation, representing the volume of the enclosed space completely exchanged by input and outlet of air.) At a height of thirty metres (100 feet) the rate of change with the same amount of opening could be treble that at ground-level. Infiltration also increases with height due to the increased wind speeds, so that it will be seen that the control of natural ventilation at higher levels is inherently a more difficult problem than at lower levels. In addition, the opening of narrow gaps for ventilation carries the risk of setting up a vibration of air, as in an organ pipe, causing whistling or moaning sounds. Therefore it is necessary to ensure that no reed-like aperture is created, and special designs of ventilator are necessary.

Orientation

The orientation of walls has been referred to as a factor in determining exposure to wind and precipitation and thus influencing heat loss. There is, however, heat gain to be considered. Heat gain varies during the day and according to the season, and produces a strong variation in the heat gain of walls of differing aspect – the low-angle morning and evening sun not providing as much heat as the southerly noonday sun near its zenith. To all intents and purposes, the sun in the British Isles is never north and, therefore, north-facing walls do not benefit from direct solar gain, except by sky-radiation or heat reflected from suitable surfaces around the building. East-facing walls receive the morning sun which gradually increases in strength up to midday, while west-facing walls gain in the afternoon. South-facing walls receive sun throughout the day in increasing, then declining strength, and at a high or low angle of incidence depending upon whether it is summer or winter.

Apart from the obvious planning implications of this, there is clearly a variation in the amount of solar heat received by walls and therefore a differing thermal differential between the inside and outside surfaces of walls and roofs, according to the aspect. Thus, irrespective of the actual air temperature, a south wall or window will not lose as much heat as one facing north in winter. In summer, the solar heat gain of a south-facing wall or window will be considerably more than that of a north face.

The windows of a room account for the larger part of the solar heat gain or loss in normal buildings, since the greater mass of the wall responds more slowly, i.e. the thermal inertia is greater. Thus, the effects of sun on walls can continue to affect the internal environment up to eight hours after the gains occur.

In other climates where the sun is more consistently shining, solar gain is dealt with as a principal design factor and results in the incorporation of sun-breakers or large overhung balconies as part of the structure, or shutters and blinds as an integral part of (or even completely replacing) windows. This solar radiation raises the temperature of the building fabric and the interior surfaces, which in turn creates the effect of high mean radiant temperatures referred to in Chapter 4, with consequent discomfort for occupants. As explained at the outset, the object of controlling the internal climate is to create comfortable living or working conditions for human beings, and the control of the radiant temperature as well as of the air temperature and humidity must be considered at the outset.

The architect is usually in the position of making basic decisions concerning the planning of buildings, their form and expression as well as their philosophy and content. He therefore makes decisions which determine the construction of the enclosure, the fenestration and the orientation – all factors which have an immediate bearing on the balance of heat – and he must therefore be fully aware of the various factors affecting the thermal characteristics of the building and the consequences of his choices if the overall performance of the building is to be acceptable.

The relevance of orientation to solar heat gain is, in itself, straightforward, but, recalling the earlier discussion of prevailing winds and precipitation, it will also be clear that orientation relates also to wind-direction. Therefore, while taking advantage of the south-westerly afternoon sun, a room also exposes itself to the wet westerly wind – and so on.

Generally speaking, the massive scale of natural phenomena makes it desirable for us to adapt the tendencies of nature to suit our purposes, rather than to try to bend them to our own requirements. So, in a west-facing room, it may be preferable to reduce the opening windows to small dimensions and put, say, the external doors on a different wall where the wind pressure is usually less. Various combinations of sun and wind arise at different seasons but, clearly, for problems of space heating alone, the winter combinations must be used in design; while for cooling and natural ventilation the summer will be the important season. Neglect of the effects of either of the two seasons can result in a poor thermal environment within the building.

Ventilation

Ventilation is not entirely a subjective matter despite the wide disparity in preferences. There is a certain basic minimum of air required for satisfactory operation of body functions, but, in most of the conventional constructions, infiltration and casual ventilation due to doors opening and shutting provides sufficient air for this purpose. In rooms with fires or other combustion heaters, infiltration may still provide sufficient air for efficient combustion due to the larger pressure-differential caused by the draught of the flue; but reliance on such casual sources of air leads to definite discomfort from draughts.

From the foregoing, it can be inferred that natural ventilation in winter is a matter of the controlled inlet of air at acceptably low velocities in such a manner that the new air will be warmed before striking an occupant of the space. In summer, it can usually be regarded as the induction of generous amounts of the warm external air so as to increase the rate of evaporation of perspiration.

To achieve this aim in summer, it is desirable so to arrange the openings that a current of air can flow in at one side of the space and out at the opposite side or through the roof. Cross-ventilation is not always feasible in complex structures, particularly in buildings of multiple occupancy such as flats or office blocks, but it is possible to secure a fair degree of air-change by the use of openings producing a 'stack effect' due to differences in wind pressures or internal-to-external thermal conditions.

Mechanical ventilation of buildings follows the same general principles as natural ventilation, but gives much more precise control of air-change. It is also more amenable to close temperature control and therefore produces the conditions required with the minimum of discomfort. In winter, the fresh air introduced is kept to a minimum while the existing air within the building is recirculated so as to keep the cost of heating the air to a minimum. In summer, unless the air is artificially cooled, the reverse applies – the maximum amount of fresh air being introduced, even to the complete exclusion of recirculation, so as to increase the cooling effect due to air movement. When the air is artificially cooled, it is economical, having treated the air, to recirculate the maximum and admit the minimum amount of fresh air compatible with ventilation requirements.

Clearly, powerful currents of air may be undesirable even in summer if, for example, paper work or paint-spraying takes place in the space; so it is preferable to minimize the velocity of a large cross-section of air and avoid high-speed concentrated streams of air to ventilate spaces used for occupation.

In the British Isles, it is rare to encounter long spells of very calm air, as the nature of our weather pattern results in fairly regular winds. Therefore, the problem of designing for natural ventilation in the summer is not unduly difficult. On the other hand, our climate includes the likelihood of frosts and cold winds between October and May and the virtual certainty of them from December to March. So the problem of designing for heating efficiency is a matter of considerable concern to the architect if economy of use is to be one of the factors in determining the ultimate form of the building. In other climates, the need for ventilation assumes greater importance than here and its influence on building form is very evident. The lofty rooms of the sub-tropical buildings of Italy and Spain, the light perforated screens of Moorish design, and the deliberate effort to site buildings, and even whole towns (Sintra, in Portugal, or the hill stations of Imperial India), on elevated sites to try to seek coolness and breezes, all point to the adoption of sites and building details to achieve an environmental aim, despite difficult local climate conditions.

Nowadays, of course, we can artificially create both our own breezes and coolness at a price, but in so doing we must aim at the maximum economy of operation which springs from properly designed buildings. Let us, therefore, look at the main causes of loss in order to see how best to control them.

The building shape

Firstly, we have seen that the construction, modified by the aspect, determines the rate of heat loss through a given area of envelope structure. Secondly, we know that glazing carries a dual penalty of high radiant loss and high radiant gain. From the first, we can infer two things: (a) that high standards of insulation reduce heat loss; (b) that the reduction of the envelope area reduces the surface from which heat can be lost. From the second, we can deduce that the area of fenestration should be controlled not only by daylighting but also by thermal considerations.

The most efficient shape for the enclosure of space by a minimum membrane is the sphere but, except for a few very specialized functions like nuclear piles and astronomy, the sphere is a difficult and dominating shape to use in buildings. It is also not an 'open-ended form', capable of easy extension. Hexagonal solids are more flexible but, like circular forms, are not amenable to convenient subdivision. The most efficient shape (in this context) in rectangular forms is the cube, and it is therefore in cubical envelopes that much building finds its expression, because not only is there an economy of enclosure-cost relative to volume enclosed, but also an economy of circulation distances within the enclosure. If vertical movement were as easy as horizontal movement, this argument might be extended to large structures; but the fact that it is not, allied with our general dependence on natural light and air, results in the evolution of extended rectangular solids or pierced cubes.

Position of plant

However, it is reasonable, on thermal efficiency grounds, to concentrate the building volume so as to reduce heat losses. This brings another factor into account – where to place the plant within the volume? Bearing in mind the economy of circulation available in compact buildings, it will be clear that to place the prime heat-source at the centre of the volume would result in short distances for services to run to any part of the building envelope. But, it is rarely convenient to site plant rooms totally within the building, if only on account of the need for access for plant removal, fuel delivery and so on.

To place the plant room at the 'centre of gravity' of the volume is, in any case, not necessarily to achieve maximum economy in distribution. The reason is that different parts of a complex building require different heat supply, either for the functions carried on in the various spaces or because of the orientation of the external wall, sometimes both. So, it will be readily seen that the decision, say, to place a cold store on the south side of a building, or a turkish bath on the north side, affects the demand for heat in two ways – by function and by orientation. A complex building, therefore, has a complicated pattern of heating or cooling load, and the distribution of heat is determined by that pattern. The centre of heat load may be regarded as the centre of gravity of heat demands, and the placing of the prime heating-plant near to this thermal centre will bring economies of distribution – not only cost economies but also of size of services, a matter of some importance where space to run services is at a premium.

Designing for economy

Another way of looking at the problem of distribution and efficiency is to so arrange the layout of the building as to ensure economy of servicing. All students will be aware of the desirability of concentrating plumbing services in buildings to achieve simplicity of supply and disposal and to facilitate the concealment of pipes. In concentrating similar functions, the aim is to make for an easier, more efficient and economical system of the service concerned. Much the same approach should be adopted with all aspects of mechanical services, and the more sophisticated the services become, the more important it is to take them into account at an early stage of planning. The creation of zones of intensity of servicing or zones of intensity of control of internal climate may not be of over-riding importance generally, but disregard of the benefits resulting from acceptance of this further discipline in design can, in buildings such as laboratories or hospitals, lead to disproportionately high cost of servicing.

Let us look, then, at some of the matters which may induce the architect to condition his thinking at an early stage.

As was mentioned earlier, solar gains affect the internal environment, reducing the heating required in winter, but increasing the cooling load in summer. While the subject of cooling is dealt with later, it is clearly desirable to place cooled spaces in locations where solar gains will be minimized. Similarly, a boiler-house might be ill-placed on a southern exposure where solar gain added to boiler heat-emission will tend to create uncomfortably hot working conditions for attendants.

Zonal problems

Buildings whose functions range over a wide field of activities demand the most consideration as to zoning, since simpler buildings can be serviced tolerably well by adjustments of a normal set of conditions. A house, for instance, requires much the same conditions throughout, irrespective of the details of the functions carried out, and can be serviced by simple forms of heating and cooling which allow only a narrow range of control. An office building may be much the same, unless sophisticated accounting machinery is installed, in which case special provision of air-conditioning for the area concerned would be considered. But a factory producing metal components, for example, may include the following incompatible functions: offices, raw-materials store, foundry, milling shop, heat-treatment plant, descaling process, acid bath, plating shop, spray-painting, electrolytic painting, enamelling, welding shop, assembly, despatch, vehicle maintenance, welfare areas such as first aid and canteen areas, and, possibly, a product control laboratory.

Some of these functions can be grouped into thermal zones by their environmental demands: for instance, foundry, heat-treatment and welding tend to be warm areas due to their process heat; spray-painting, plating and descaling tend to be high ventilation areas; raw-materials stores and despatch tend to be subject to fluctuations of temperature due to their relationship to delivery of goods and the need to open external doors in order to function. The concentration of these groups in the physical planning of the factory will naturally produce areas of high demand, and the relationship of such areas to each other will determine the positioning of primary heating/cooling plant, and the location of any secondary plant serving particular zones.

Time zoning

Another prime consideration of zoning is the time factor. Certain functions in a complex building may run continuously while others are intermittent. Taking the factory again as an example: foundries usually run continuously as the warm-up time for furnaces dictates that they never be shut down; offices, on the other hand, will only operate during eight or nine hours of the day. With this disparity of requirement, which may operate in various ways through the range of functions, the services plant must be so sub-divided that a departmental control over heating and ventilation can be achieved. This may therefore result in another discipline of design which groups together the functions according to their programme requirements.

7: Thermal Balance

BEFORE GOING INTO FURTHER DETAILS of the means of providing for the comfort needs of man in buildings, it is desirable, once again, to state the aims of servicing and the human factors involved:

Basically, the aim is to design buildings, efficient in terms of thermal properties and economical in the provision of engineering services, to provide heating or cooling to maintain the body at a nearly constant temperature.

The body stores energy derived from food and uses some of the energy to do work. It also generates heat to maintain body functions and at the same time dissipates heat to its surroundings in three ways:

(a)	Radiation	45 per cent	at 18°C (65°F) approximately
(b)	Convection	30 per cent	
(c)	Evaporation	25 per cent	

Conduction losses can be discounted as a serious cause of heat loss, since the area of direct contact is usually very small.

Discomfort arises when these body functions are unable to maintain the normal body-temperature, and it is therefore necessary so to balance the environmental systems as to provide the conditions in which the body can produce and dissipate the correct amount of heat. Thus, it is essential that human beings dissipate heat at a controlled rate at all times.

The four main external factors which influence the sensation of comfort are:

(a) Air temperature;
(b) Radiant temperature;
(c) Humidity;
(d) Air movement.

These factors must therefore be considered at an early design stage in the context of the local external climate, the function of the space and its location within the building.

The purpose of the building demands its own degree of comfort. For example, the church, the classroom, the factory and the hospital have differing requirements, and each will be constructed differently and therefore have different thermal characteristics.

Consequently, it is essential for the architect to have a thorough grasp of fundamentals to enable him to provide the overall environmental conditions in which the heating and mechanical services can be economically and effectively applied, resulting in an internal thermal environment for each space or element closely related to the needs of that space in terms of human comfort. The observations which follow illustrate the importance of this.

Air temperature

From various authoritative sources it can be said that the optimum air temperature in the U.K. is 18·3°C (65°F), but the range of requirement for comfort could vary from 15·6°C to 20°C (60° to 68°F) for light factory-work depending upon the degree of comfort required. But, in sedentary occupations, the range might be 19·4°C to 22·8°C (67° to 73°F), while for heavy work 12·8°C to 15·6°C (55° to 60°F) could be considered acceptable. All these temperatures are of course dry-bulb temperatures.

The architect usually has little or no control over the resultant air temperature within a building other than by designing an inadequately insulated or ventilated building, or by sacrificing the correct kind of system for something cheaper without appreciating the consequences.

Radiant temperature

Acceptable standards of heat due to radiant effect are more difficult to define but can be regarded as satisfied when the M.R.T. (mean radiant temperature) ranges from 16·7°C to 20°C (62° to 68°F) depending upon occupational requirements.

It is important to remember that heat always travels from a high-temperature source to a low-temperature source, never the reverse; in consequence, the effects of cold surface, with resultant excessive heat loss from the body to say a window or a wall, are to be avoided (hence the normal location of heating units). Similarly, the solar effect (i.e. the direct radiant gain from the sun) through a window usually proves quite unacceptable to the unfortunate occupants in the path of such heat gain, hence the provision of blinds or balconies or mullions, the first usually resulting from lack of appreciation of sun-effect in terms of radiant heat.

The architect's decisions can influence the degree of M.R.T. *through his choice of building construction, orientation and internal finishes.*

Humidity

Comfort is achieved over a fairly wide range of forty to sixty-five per cent R.H., which is acceptable within the ordinary dry-bulb temperature range; however, wide variations outside the normal comfort temperature are to be avoided. (The combination of dry-bulb and wet-bulb determine the relative humidity of the air. This in turn will indicate the conditions at which condensation will occur on cold surfaces, for example.)

The architect's decisions will not influence the R.H., *but his choice of materials can produce bad results from ignorance of the effects produced by varying* R.H. *values at different temperatures: i.e. cold ceilings or floors are often a source of condensation, as are single-glazed windows in locations of high humidity.*

Air movement

Subjective experiments indicate that the optimum value for air movement can be taken as about 0·17 m/sec (30 ft/min), movement being just perceptible. Speeds in excess of 0·5 m/sec (100 ft/min) are considered 'very draughty', and, below 0·1 m/sec (20 ft/min), are considered 'airless'. However, the effects of air movement are complex and differ with circumstances, but they are particularly important if the method of providing comfort is an all-air system for either heating or cooling.

The architect's decisions can and will influence air movement by the nature of his acceptance or otherwise of grille positions, obstructions to air flow caused by beams, too high or too low a ceiling, remoteness of discharge from sources of maximum heat loss or heat gain. To achieve the best results for the occupants, close co-operation and inventiveness at the planning stage are essential.

The combination of these four factors in ways which produce satisfactory environments for various occupational purposes is a complicated procedure, but subjective tests over many years have resulted in the production of charts and nomographs by H.M.S.O. and other authorities, by the use of which it is possible to arrive at suitable levels of comfort conditions for these varying factors, according to geographical locations within the building and the type of occupancy. It is therefore possible, by the use of these graphic aids, to establish the degree of warming or cooling which will be acceptable. To do this, a number of instruments producing standard indices have been devised, and are briefly described below.

Measurement of thermal comfort

The following instruments are used to ascertain comfort conditions:

1. Ordinary mercury-in-glass thermometer – measures the air temperature and is referred to as the *dry-bulb temperature.*
2. Thermometer similar to the above, but having the bulb covered with a wet muslin; it gives the *wet-bulb temperature.*

   These two are often combined into one instrument (like a rattle) which is whirled round by hand and the resultant readings on the two thermometers give the wet-bulb and dry-bulb temperatures. From tables of hygrometric data, the resultant relative humidity can be ascertained.
3. An ordinary thermometer sealed into a blackened copper sphere (about 6 in dia.), referred to as a *Globe Thermometer,* which assists in finding the mean radiant temperature.
4. A mercury-in-glass thermometer, having a larger bulb of a specific size, is subjected to test, and a 'cooling factor' (due to air movement) is determined. This factor is then applied to readings taken from the thermometer – known as a *Katathermometer* – and will indicate air movement down to 0·05 m/sec (10 ft/min) and upwards of 50 m/sec (1000 ft/min). Various anemometers, electrical and mechanical, are used to ascertain air movement, particularly in systems using air as the heating or cooling medium.

The use of a combination of these instruments in a number of subjective tests has resulted in the production of nomographs, from which can be ascertained the probable comfort range a given set of conditions will produce. There are also several aids to suitable design in the scales and indices following:

Scales of Warmth or Comfort

  (i) *Air temperature* – Simple to determine and, within the normal range of temperatures, air movement etc., quite adequate.

  (ii) *The Normal Corrected Effective Temperature Scale (n.c.e.t.)* – See Chart 4, *Appendix* A.

  This is used when there is:

  (*a*) extreme temperature with extreme air speed;

  (*b*) strong interchange of radiation or convection between people and their environment;

  (*c*) an environment inducing perspiration.

  (iii) *The Wet-bulb Globe-Thermometer Index*

  This scale of comfort is used in warmer climates than is experienced in the U.K.

  (iv) *The Equatorial Comfort Index*

  Preferably used in warm humid climates for temperatures above 24°C (75°F) effective temperature.

The above indices serve to indicate the average conditions required to enable the normal body reactions to reach equilibrium with the environment experienced.

It must therefore be a pre-requisite in the planning of a building to aim to provide the 'right' environment to which the normal reactions of the body can adjust themselves.

Elementary physics relating to comfort

Earlier we ascertained the four principal factors affecting comfort as:

(*a*) Air temperature;

(*b*) Radiant temperature;

(*c*) Humidity;

(*d*) Air movement.

We also established the four means of heat transfer between the human body and the environment as:

(*a*) Radiation;

(*b*) Convection;

(*c*) Evaporative cooling;
(*d*) Conduction.

Radiation, as we have said, behaves like light. All materials radiate heat in direct ratio to their temperature, thus a cool body and a hot body both radiate heat to each other, but the net gain of heat occurs in the cool body while the net loss of heat occurs in the hot body. The rate at which a body can dissipate heat (i.e. its *emissivity*) or absorb heat (i.e. its *absorptivity*) also depends upon its colour and the quality of the surface. Bright metallic surfaces have low emissivity and absorptivity but reflect radiant heat well. Matt surfaces have greater absorptivity and emissivity. Dark colours have high absorptivity and emissivity. Hence, to reduce radiant-heat intake on a sunny day, we wear light-coloured clothes, and astronauts wear metallic-finished clothes to reflect away the powerful radiation in space. Radiation is virtually unaffected by its passage through **air** and it does not materially affect the temperature of the air.

Convection currents, causing the transfer of heat, arise from all warm bodies; they also result from variations in pressure, such as winds. Thus, there are three principal convective gains and losses in buildings:

(i) Heat loss due to infiltration in winter.
(ii) Heat gain due to infiltration in summer.
(iii) Convection air currents passing over surfaces at higher or lower temperature, the air thereby gaining or losing heat.

The human body also loses or gains heat from air movement over its surface.

Evaporative cooling occurs as a result of the latent heat released or absorbed when a change of state (e.g. from water to steam or the reverse) takes place. When water evaporates, heat is absorbed from its surroundings, so the evaporation of per-spiration causes a drop in skin temperature. Similarly, a wet porous building material loses heat as evaporation takes place.

Conduction, although it is the principal cause of heat loss from buildings, is of little account in bodily heat loss since the surface area of contact whereby heat can be lost is usually small.

The combination of these four methods of heat transfer, and the resultant environmental situation produced, reacts upon the human body and produces sensations of comfort or discomfort.

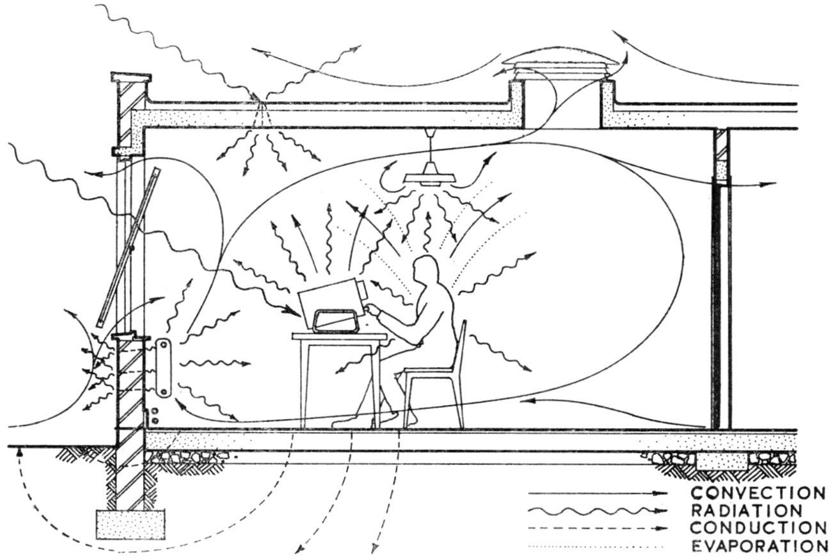

FIG. 7.1. The Human Body in its Surroundings.

Fig. 7.1 shows the human body in its surroundings with the manifold inter-actions of energy which take place. It will be noted that some of the heat-transfer paths may use a sequence of several of the methods described above.

Thermal exchange between the body and its surroundings

While the body temperature remains constant, the heat produced within the body is equal to the body heat loss: and the body heat-production (M) equals the total energy-production (metabolism).

The equation can be expressed:

$$M - W = E + R + CV + S.$$

where W = rate expended in mechanical work.

E, R and CV = rate of heat loss by evaporation, radiation and convection.

S = rate of heat being stored in the body.

The thermal transference between a person and his environment is shown in Fig. 7.1.

The establishment of valid figures for these various factors is still a matter of research but, since the equation varies as the work and metabolism vary, it is only possible to work on average values as a basis suitable for the type of occupancy.

Reference to the graph (Chart 2, *Appendix* A) indicates the different rates of energy-dissipation in terms of latent and sensible heat given off from the body in various states of rest and work. Such information relates to the environmental needs of human beings; but how to provide the conditions within the building is the problem the architect must assist in solving.

Taking into account all the various aspects of heat transfer, the basic equations for thermal comfort of a space can be written as follows:

Winter Season

$$\boxed{\text{Heat Loss}} \quad = \quad \boxed{\text{Heat Gain}}$$

That is:

Fabric losses from: Roof ⎱ Walls ⎰ (dependent on insulation) Floor ⎰ Windows (radiant, instantaneous); (conducted, delayed). Air changes, losses from: Ventilation (a) natural; (b) artificial. Infiltration	=	Gains from: Personnel Lighting Machinery Solar (with exposed external surfaces this results in a variable gain and, except in very mild weather, provides insufficient heat. Completely internal spaces may gain more than enough heat from the first three sources and consequently may need to be cooled, even in winter). Heating (with controls to adjust for variable heat gains).

(N.B. It must be realized that, though in normal buildings the solar heat gain is inadequate for full heating, special buildings of massive construction can be designed into which solar heat, collected by a special south-facing wall, is directed. A recent school has been heated by this method, supplemented by the heat gains from lighting and personnel, and though a conventional heating system was installed as a precaution it has not yet had to be used. Such special construction relies upon highly efficient insulation and high thermal inertia. The economics of the system have not yet been released.)

Summer Season

| Heat Gain | = | Heat Loss |

That is:

<table>
<tr><td>

Gains from:
Solar (radiant through glass
    instantaneous);
    (conducted through
    fabric, therefore
    delayed).

Lighting

Machinery

Personnel
  (sensible heat)
  (latent heat from moisture
  in air).

</td><td>=</td><td>

Losses from:
Air movement
    (latent heat of evaporation
    of moisture);
    (sensible heat by convec-
    tion).
Hence the value of measures
designed to reduce heat
gain:
  (1) shade (against solar
    gain through glass).
  (2) insulation (against
    solar gain conducted
    through fabric).
These can, in temperate
climates, keep conditions
within tolerable limits, e.g.
the scale of Normal Cor-
rected Effective Tempera-
ture. (Chart 4, *Appendix* A.)
Cooling, artificially imposed
(with controls to adjust for
variable heat gains).

</td></tr>
</table>

Comparison of these two equations shows a number of common factors as well as some major differences. Firstly, the importance of insulation against heat losses for winter is balanced by the importance of such insulation against heat gains in summer, but, whereas fabric losses predominate in winter, solar gains may provide the greater problem in summer.

Secondly, ventilation losses have similar importance in both winter and summer: in one case as a loss to be optimized, and in the other as a means of cooling to be taken advantage of.

Thirdly, neither equation balances without the imposition of additional heat gain or loss by heating or cooling, except in favourable conditions; and this additional heat transfer must be adjustable because of the variability of the naturally occurring heat transfer. Ideally, the adjustment should be controlled by the environment itself, i.e. thermostatically.

Note that the location of an internal space or zone within the building may demand opposite treatment from the remainder of the building, owing to its insulation from summer heat loss and the corresponding trapped heat from machines, lighting and personnel.

The balanced environment

The balanced environment is one which provides the correct level of comfort for the function of the space, and the determination of the level can result from statutorily imposed standards (as a minimum), recommended or desired standards, or from research into the particular problems of specialized functions. Much of the information is tabulated, mainly in terms of air temperature and air movement (air change) alone, though humidity is often an important factor in electronic work and is vital in textiles. Similarly, the balanced environment takes account of the time element arising from the pattern of working hours or occupation of the building, and these various requirements make it desirable for the designers of the building to ask themselves basic questions at the outset. Some of these questions might be:

(a) What thermal inertia is desirable in the building bearing in mind its pattern of occupation?

(b) Will the selection of a low thermal inertia (resulting from lightweight building) give rise to noise nuisance from a busy road?

(c) Does the ventilation rate (required for some special process) demand a high degree of radiant heating?

(d) Will the selection of a wide but shallow heating zone (arising from the function of the building) produce internal zones which are isolated?

And so on. . . .

Meeting the thermal requirements of the varied functions of buildings is achieved in two ways. Firstly, the basic concept of the building (including, as it should, consideration of thermal needs) determines to a large degree the thermal characteristics of the building fabric, when subjected to the local climate. Secondly, the mechanical systems, devised by the heating engineer, have to produce the pre-determined comfort conditions related to each function or functional zone.

Clearly, the first affects the second, making it difficult or straightforward, costly or economical. Thus, the architect should consider how he can construct the building so as to extract the maximum benefit from the envelope. He therefore considers the insulation value of his external membranes, the rate of ventilation resulting from his disposition or design of openings, and the radiant temperature resulting from his finishes.

Control of the radiant temperature is achieved by the use of suitable surfaces in the space or zone. It will be recalled that materials were described as possessing varying emissivities according to their nature or, more particularly, their surface. Polished foil has a high reflective factor but low emissivity, while a matt dark surface has high emissivity. To maintain a mean radiant temperature as close to the actual air temperature as possible requires the selection of finishes which, given the level of heating required, will emit radiant heat in the quantity required. If the insulation value of the building envelope is low, the temperature of the inner surface is low, thus the U value of a wall relates not only to the rate of conducted heat loss but also to the level of radiant heat emission of the inner surface.

Tabulated below are examples of the effect of U values on the temperature of the inner surface of a building enclosure, when the external-air temperature is $-1°C$ ($30°F$), and the internal-air temperature is $15·5°C$ ($60°F$).

External Air Temperature (to)		Trans-mittance (U)		Internal Air Temperature (ti)		Temperature of Inner Surface (ts)		Temperature Difference Air-to-Surface	
C	F	S.I.	Imp.	C	F	C	F	C	F
—1	30°	5·8	1·0	15·5°	60°	4°	39°	11·5°	21°
,,	,,	2·9	0·5	,,	,,	10°	49·5°	5·5°	10·5°
,,	,,	1·9	0·33	,,	,,	12°	53°	3·5°	7°
,,	,,	1·1	0·2	,,	,,	13°	55·5°	2·5°	4·5°
,,	,,	0·6	0·1	,,	,,	14·5°	58°	1·0°	2°

Fig. 7.2.   Effect of Transmittance on Surface Temperature.

The raising of surface temperature can allow a reduced air temperature approaching the optimum value required for comfort. With intermittent heating, a rapid rate of rise of surface temperature is desirable so as rapidly to create comfort conditions. To do this, the mass of construction within the main insulant should be as small as possible; in fact the best arrange-

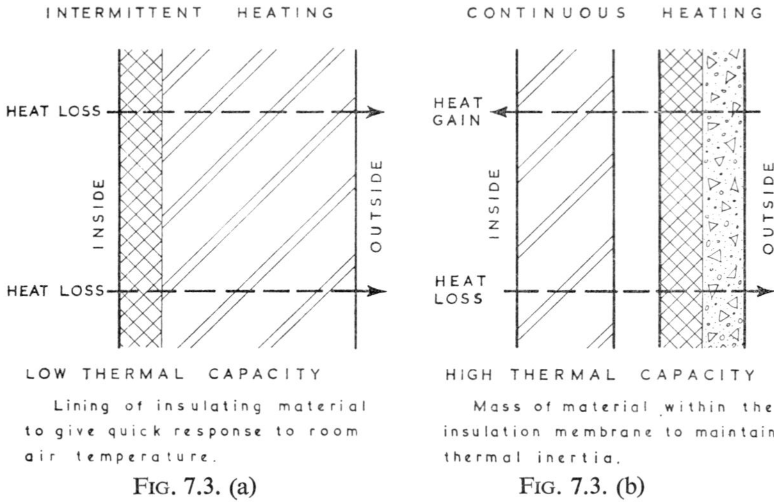

FIG. 7.3. (a)   FIG. 7.3. (b)

The Effects of the location of Thermal Insulation.

ment is to apply the insulant to the inner surface of the construction and have virtually no inert mass to heat (see Fig. 7.3 (a)). Briefly, the rate of rise of surface temperature is inversely proportional to the thermal conductivity and the thermal capacity. For continuous heating, therefore, the position of the insulant is less important but is preferably near the outside so as to maintain the mass of the structure at the higher indoor temperature (see Fig. 7.3 (b)).

This argument applies equally to walls, roof and floor. Floors with low thermal capacity and conductivity, such as wood, cork, carpets, etc., have a surface temperature close to the air temperature. Floor finishes of tile or plastic laid direct on concrete have high conductivity and capacity and, therefore, their surface temperature rises slowly compared with that of the air. This can result in the low surface temperature being below dewpoint, giving rise to condensation on the cold surface.

A simple example of heat transfer calculation

From what has gone before, and by referring to information tabulated in *Appendix* B, it is now possible for a simple problem of heat transfer to be considered.

Let us consider a small single-storey house of traditional construction, to be used for week-end occupation only, and let us suppose that it is sited in an inland area of average exposure, the soil being a gravel-clay mixture of moderate water content. Such a dwelling will tend to resolve into two heating zones – living and sleeping – each with different timing and temperature requirements.

As the occupation is intermittent, a rapid warm-up is necessary and no great degree of thermal inertia is required. To provide flexibility of use by the occupants, the living areas are to be open-planned. The house is not in London, so the *Building Regulations, 1965*, apply.

From these facts certain basic decisions should be made:

1. No extraordinary measures need be considered on account of exposure or soil.

2. The heating system could operate so as to provide overall background heating with the facility of boosting either zone to a higher level, but not necessarily both at once. The living zone would normally be at a temperature of about 19°C (65°F) and the sleeping zone at 13°C (55°F), with minimum 3°C (5°F) below.

3. A time-switch system of control which will pre-heat the house before arrival will be desirable.

4. The lack of occupation during the week necessitates some automatic protection against frost damage to water-systems (unless these are drained down).

5. The rapid warm-up and low thermal inertia requires that the construction be either:
   (*a*) Lightweight,
   (*b*) Insulated at the inner surface to keep the main mass of the construction outside the thermal envelope.

6. The Building Regulations minimum conductances are:

$$\text{Walls: } U = 1 \cdot 71 \ \frac{\text{W}}{\text{m}^2 \text{ deg C}} \ (0 \cdot 30 \ \frac{\text{Btu}}{\text{ft}^2 \text{ hr deg F}})$$

$$\text{Floor: } U = 1 \cdot 42 \quad , \quad (0 \cdot 25 \quad , \quad)$$

$$\text{Roof: } U = 1 \cdot 42 \quad , \quad (0 \cdot 25 \quad , \quad)$$

7. Recommended conductances for economical heating and a comfortable thermal environment are:

Walls – living room: $U = 0.86$ to 1.14 (0.15 to 0.2)
  – other rooms: 1.14 to 1.71 (0.2 to 0.3)
Floor: 0.86 (0.15)
Glazing – south aspect: 4.56 (0.8)
    – north aspect: 5.7 (1.0)
Roof (or roof and ceiling
  combined): 1.14 to 1.71 (0.2 to 0.3)

Let us assume that the house is 10 metres square and has windows on each wall amounting to 30 per cent of the wall surface. The floor area is divided equally into living and sleeping areas, the construction generally being:

Walls – living area:
    0.28 m (11 in) cavity wall lined inside with 12 mm ($\frac{1}{2}$ in) fibreboard (Class I spread of flame) on 25 mm (1 in) battens.

   „   – sleeping area:
    0.28 m (11 in) cavity wall with inner skin of solid clinker blocks.

Floor: 0.15 m (6 in) concrete with cork tile finish.

Glazing: sealed double-glass units.

Roof: 20 mm ($\frac{3}{4}$ in) asphalt on 25 mm (1 in) cork-board on 25 mm (1 in) chipboard on timber joists with 10 mm ($\frac{3}{8}$ in) plasterboard ceiling with 3 mm ($\frac{1}{8}$ in) skim coat.

The heat loss calculations may be presented as follows (figures in brackets are approximate Imperial equivalents):

Particulars

Size of dwelling: 10 m $\times$ 10 m $=$ 100 m² (1076 sq. ft)
Height: 2·5 m (8·2 ft)
Area of walls (gross): $4 \times 10 \times 2.5$ $=$ 100 m² (1076)
Area of windows: $30\% \times 100$ $=$ 30 m² (323)

Heat Loss – Walls

The enclosure of the living-zone, therefore, is an area of 50 m² with 15 m² of window leaving 35 m² of wall. Taking $U = 1.09$ (0.19) and a temperature difference (external air, $-1°$C (30°F); internal air, 19°C (65°F)) of 20°C (35°F) the heat loss from the wall is as follows:

$$35 \times 1.09 \times 20 = 763 \text{ watts}$$
$$(376 \times 0.19 \times 35 = 2500 \text{ Btu/hr})$$

The sleeping-zone walls are of a similar area, but $U = 1\cdot37$ (0·24) and the temperature difference is 14°C (25°F). The heat loss is therefore as follows:

$$35 \times 1\cdot37 \times 14 \;=\; 671 \text{ watts}$$
$$(376 \times 0\cdot24 \times 25 \;=\; 2256 \text{ Btu/hr})$$

Heat Loss – Windows

The total area of the windows is 30 m², and comprises the windows in the living-zone which faces to the south and those in the sleeping-zone which faces to the north.

The heat loss for the windows varies because of the varying temperature difference, though the conductances are the same. Thus:

Living-zone heat loss =
$$15 \times 2\cdot86 \times 20 \;=\; 858 \text{ watts}$$
$$(161 \times 0\cdot5 \;\times 35 \;=\; 2817 \text{ Btu/hr})$$
Sleeping-zone heat loss =
$$15 \times 2\cdot86 \times 14 \;=\; 600 \text{ watts}$$
$$(161 \times 0\cdot5 \;\times 25 \;=\; 2012 \text{ Btu/hr})$$

Heat Loss – Floor

The conductance of solid floors varies over the area of the floor, there being a higher loss at the edges. Consequently the figures usually tabulated represent the average rate of conductance.

Floor area of living-zone = 50 m²
Heat loss $\quad 50 \times 0\cdot86 \times 20 \;=\; 860$ watts
$\qquad (538 \times 0\cdot15 \times 35 \;=\; 2824$ Btu/hr)
Floor area of sleeping-zone = 50 m²
Heat loss $\quad 50 \times 0\cdot86 \times 14 \;=\; 602$ watts
$\qquad (538 \times 0\cdot15 \times 25 \;=\; 2017$ Btu/hr)

Heat Loss – Roof

Roof area of living-zone = 50 m²
Heat loss $\quad 50 \times 1\cdot43 \times 20 \;=\; 1430$ watts
$\qquad (538 \times 0\cdot25 \times 35 \;=\; 4707$ Btu/hr)
Roof area of sleeping-zone = 50 m²
Heat loss $\quad 50 \times 1\cdot43 \times 14 \;=\; 1001$ watts
$\qquad (538 \times 0\cdot25 \times 25 \;=\; 3362$ Btu/hr)

Total Conducted Heat Loss of Building:

$$763 + 671 + 858 + 600 + 860 + 602 + 1430 + 1001$$
$$= 6785 \text{ watts}$$
$$(2500 + 2256 + 2817 + 2012 + 2824 + 2017 + 4707 + 3362$$
$$= 22{,}495 \text{ Btu/hr})$$

Ventilation

The introduction of air (due to infiltration and natural ventilation) imposes a further heat loss, as the warm air is lost and cool air takes its place. In our heat-loss calculation, therefore, we must allow for the heating of this new air to the design temperature. For domestic buildings, this air change can be taken as two complete changes of air every hour, and the heat required to raise this air by one degree celsius is $1 \cdot 2$ kJ/m³ (0·02 Btu/cu. ft/°F). Thus we must calculate the cubic volume of air (for each temperature difference) and multiply it by the number of air changes per hour.

Heat loss, living-zone: Volume $= \dfrac{100 \times 2 \cdot 5}{2} = 125$ m³

@ 2 air changes per hour $= \dfrac{125 \times 2}{3600} = 0 \cdot 07$ m³/sec

@ 1200 J/m³ °C $= 0 \cdot 07 \times 1200 = 84$ W/°C

@ 20°C temperature difference $= 84 \times 20 \quad = 1680$ watts

$(\frac{1}{2} (1076 \times 8 \cdot 2 \times 2 \times 0 \cdot 02 \times 35) \quad = 6176$ Btu/hr)

Heat loss, sleeping-zone: Volume $= \dfrac{100 \times 2 \cdot 5}{2} = 125$ m³

@ 2 air changes per hour $= \dfrac{125 \times 2}{3600} = 0 \cdot 07$ m³/sec

@ 1200 J/m³°C $= 0 \cdot 07 \times 1200 = 84$ W/°C

@ 14°C temperature difference $= 84 \times 14 = 1176$ watts

$(\frac{1}{2}(1076 \times 8 \cdot 2 \times 2 \times 0 \cdot 02 \times 25) \quad = 4411$ Btu/hr)

Thus, the total heat loss of the building, conducted and due to ventilation, is:

$$6785 + 1680 + 1176 = \quad 9641 \text{ watts}$$
$$(22{,}495 + 6176 + 4411 = 33{,}082 \text{ Btu/hr})$$

It follows, therefore, that to offset the normal heat losses for the desired temperatures, approximately 9600 watts (33,000 Btu/hr) must be provided. This can be considered as the net heat requirement of the building, but, depending upon the thermal characteristics of the structure and the necessity for intermittent or continuous occupation, a margin of approximately 25 per cent over and above the net figure would be allowed so as to be on the safe side.

Therefore

$$9600 + 2400 = 12,000 \text{ watts} = 12 \text{ kW}$$

$$(33,000 + 8250 = 41,250 \text{ Btu/hr})$$

Note that the above calculations do not include for domestic hot-water supplies which, if heated from the same primary unit, would increase the required output capacity of the boiler still further.

8: Condensation

CONDENSATION HAS BEEN TOUCHED ON earlier as a phenomenon connected with humidity. Apart from indicating a failure to appreciate the importance of surface temperature, it is unsightly and can be damaging to finishes. In unventilated spaces it can assist in producing conditions suitable for the propagation of mould and fungus, and it should therefore be prevented as far as possible.

Surface condensation

Surface condensation, resulting from the temperature of the surface being below the dewpoint of the air, occurs in two conditions, 'permanent' condensation and 'temporary' condensation. 'Permanent' condensation arises if the rate of thermal transmittance is high, causing an inherently low surface temperature, and the humidity of the space is high (as in cotton mills, jam factories, etc.). 'Temporary' condensation occurs when a sudden change in air humidity occurs in or around a cold structure – for example, when a 'warm front' succeeds a spell of cool dry weather. Similarly, artificial changes in internal atmospheric conditions, owing to cooking or laundry activities, or to social gatherings of numerous persons in confined space, produce the same effects.

Permanent condensation can be avoided only by raising the surface temperature above dewpoint by the use of insulation in the structure, or cured by providing a surface finish with good thermal insulation. Increasing the ventilation rate to reduce the build-up of moisture in the air also reduces the degree of the condition.

Temporary condensation is not normally a serious problem and is usually dealt with by ventilation, or by the application of 'anti-condensation' paints, some of which take the form of absorbent coatings capable of taking up moisture during condensation-conditions and releasing it when the humidity falls or the structure warms up. Wallpapers can act in the same way, but unsightly bubbling or pattern-staining can result.

Permanent condensation occurs on all types of surface, whether walls, ceilings or floors, if the conductance is high (i.e. if U is high). Walls and insulated ceilings normally have a fairly constant conductance over the whole of their area and, consequently, condensation appears uniformly over the surface; but solid floors have varying conductance according to the distance from the edges of the slab, and it is therefore not uncommon for the periphery of the slab to have a surface temperature below dewpoint while the remainder is above. This situation can arise because, though the temperature of the soil may be taken as constant at about 10°C (50°F) at a depth of 1 metre, there is a loss from the edge of the slab to the cooler soil above this level. In addition, the effects of rain and frost on the conductivity of the soil produce variable rates of heat loss at edges.

Consequently, with solid floors, a fringe of condensation, beneath loose coverings such as linoleum or carpets, can occur and cause damage to the floor finishes and skirtings. Fig. 8.1 shows such a situation, where edge insulation has been applied

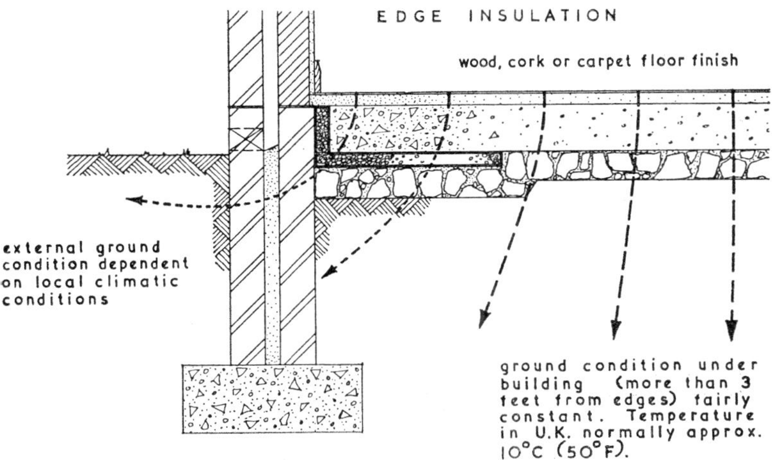

FIG. 8.1. Edge Insulation of Solid Floors.

to reduce the edge loss and maintain high surface temperature. Owing to the inherently damp condition of this detail, the insulation is usually a damp-resistant material such as expanded polystyrene or corkboard.

Condensation within structures

So far, we have considered cases where the surface temperature is below dewpoint, but it will be realized that porous materials may contain moisture and that the temperature gradient of a structure may be such that dewpoint can be reached within the thickness of the structure. In such conditions the vapour within the cells (or interstices) of the material will condense. This in itself can cause the breakdown of some materials, particularly if the condensation occurs near the outer surface where subsequent frost-action causes expansion and rupture of the cell walls. Nearer the inner surface, this frost-rupture is unlikely to occur, but the constant wetting of acidic materials like breeze concretes can cause the transmigration of chemicals to the inner surface, causing breakdown or staining of finishes. Salts in mortars and plasters also erupt on the inner surface, as efflorescence.

To prevent interstitial condensation, it is necessary to prevent

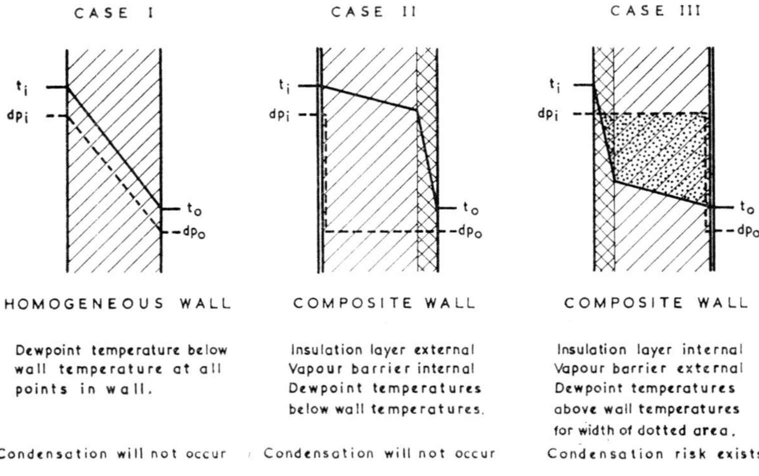

Fig. 8.2. Condensation within Walls.

the passage of moisture from the space into the structure and so to arrange the conductivities of a non-homogeneous structure that the temperature gradient is always above the dewpoint gradient. Fig. 8.2 shows three cases of walls, in two of which interstitial condensation will not occur, while in the remaining case it will occur. The actual temperature-line is shown solid and the dewpoint temperature-line is dotted.

Temperature gradients

Though temperature gradients were mentioned in Chapter 4, the calculation of the gradient through a structure has remained unexplained. However, as will be evident, the gradient depends upon the nature of the constituent materials of the structure and upon the external and internal temperatures. Given an internal temperature of 19°C (say 65°F) and external temperature of −1°C (30°F), the temperature difference is 20°C (35°F). With a non-homogeneous wall, the gradient within the wall consists of a series of straight lines at different slopes. Fig. 8.3 shows the wall with the actual temperatures at each surface plotted on the gradient. With a homogeneous wall, such as solid brickwork,

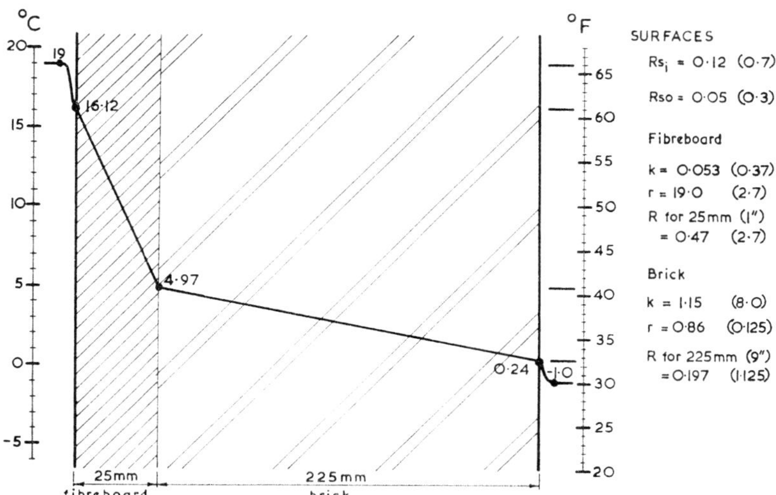

FIG. 8.3.   Thermal Gradient of a Composite
(Non-Homogeneous) Wall.

the gradient follows a straight sloping line from surface to surface, and at each surface there is a change of gradient due to surface resistance (see Fig. 8.4). The same surface temperatures are still arrived at in the case of both walls.

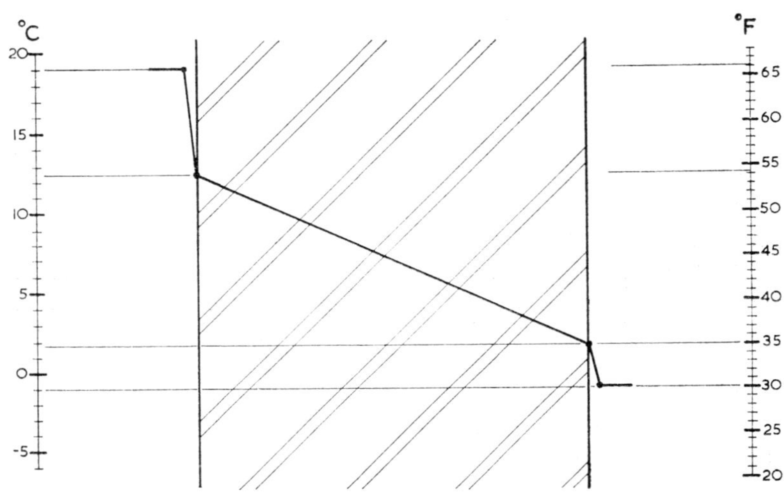

FIG. 8.4. Thermal Gradient of a Homogeneous Wall.

(N.B. The Fahrenheit temperatures do not correspond exactly with the Celsius because 19°C = 66°F).

The gradient shown on Fig. 8.3 is calculated as follows:
Total thermal resistance of wall (R_T)

$$= Rsi + \frac{L1}{k_1} + \frac{L2}{k_2} + Rso$$

$$= 0{\cdot}121 + \frac{0{\cdot}025}{0{\cdot}053} + \frac{0{\cdot}225}{1{\cdot}15} + 0{\cdot}052$$

$$= 0{\cdot}121 + 0{\cdot}47 + 0{\cdot}197 + 0{\cdot}052$$

$$= 0{\cdot}840 \qquad\qquad R_T$$

Total temper-
ature difference
across wall $\qquad = 20° \qquad\qquad$ td

Temperature Gradient (S.I. Units)

(i) Temperature at inner face of fibreboard

=Temperature of inside air

minus $\left(\dfrac{td}{R_T} \times \text{Surface Resistance}\right)$

$= 19° - \left(\dfrac{20}{0.84} \times 0.121\right)$

$= 19° - 2.88°$

$= 16.12°C$

(ii) Temperature at outer face of fibreboard
(equals temperature at inner face of brick)
=Temperature at inner face

of fibreboard minus $\left(\dfrac{td}{R_T} \times \text{Material Resistance}\right)$

$= 16.12° - \left(\dfrac{20}{0.84} \times 0.47\right)$

$= 16.12° - 11.2°$

$= 4.92°C$

(iii) Temperature at outer face of brick
=Temperature at outer face

of cork minus $\left(\dfrac{td}{R_T} \times \text{Material Resistance}\right)$

$= 4.92° - \left(\dfrac{20}{0.84} \times 0.197\right)$

$= 4.92° - 4.68°$

$= 0.24°C$

(iv) Temperature drop at outer surface of brick

$= \dfrac{td}{R_T} \times \text{Surface Resistance}$

$= \dfrac{20}{0.84} \times 0.052$

$= 1.24$

For the homogeneous wall (Fig. 8.4) the calculation is:

$R_T = 0.121 + 0.197 + 0.052 = 0.370$

td $= 20°$

(i) Temperature of inner surface of brick

$$= 19 - \left(\frac{20}{0.37} \times 0.121\right) = 19 - 6.54 = 12.46°C$$

(ii) Temperature of outer surface of brick

$$= 12.46 - \left(\frac{20}{0.37} \times 0.197\right) = 12.46 - 10.65 = 1.81°C$$

(iii) Temperature drop at outer surface

$$= \frac{20}{0.37} \times 0.052 = 2.81°$$

N.B. (1). Comparison may be made between the inside air temperature and inner surface temperatures of the two walls:
Composite wall $19°C - 16.12°C = 2.88°$ difference
Homogeneous wall $19°C - 12.46°C = 6.54°$ difference
This demonstrates the relationships shown in Fig. 7.2.
N.B. (2). In calculating R_T we have found the reciprocal of the U value of these walls.

Thus: U of the composite wall $= \dfrac{1}{0.84} = 1.19 \text{ W/m}^2 °C$

U of the homogeneous wall $= \dfrac{1}{0.370} = 2.70 \text{ W/m}^2 °C$

Temperature Gradient (Imperial Units)
Parallel calculations in imperial units for the same two structures follow a similar pattern:
(*a*) *Composite Wall* (Fig. 8.3)

$$R_T = Rsi + \frac{L1}{k1} + \frac{L2}{k2} + Rso$$

$$= 0.7 + \frac{1}{0.37} + \frac{9}{8.0} + 0.3 = 4.825 \text{ ft}^2 \text{ h } °F/Btu$$

td $= 65° - 30° \qquad\qquad = 35°F$

Gradient:

(i) Temperature at inner face of fibreboard

$$= t_i - \left(\frac{td}{R_T} \times \text{Surface resistance} \right)$$

$$= 65 - \left(\frac{35}{4 \cdot 825} \times 0 \cdot 7 \right)$$

$$= 65 - 5 \cdot 07 \qquad = 59 \cdot 93°F$$

(ii) Temperature at outer face of fibreboard

$$= 59 \cdot 93 - \left(\frac{35}{4 \cdot 825} \times 2 \cdot 7 \right)$$

$$= 59 \cdot 93 - 19 \cdot 60 \qquad = 40 \cdot 33°F$$

(iii) Temperature at outer face of brick

$$= 40 \cdot 33 - \left(\frac{35}{4 \cdot 825} \times 9 \times 0 \cdot 125 \right)$$

$$= 40 \cdot 33 - 8 \cdot 16 \qquad = 32 \cdot 17°F$$

(iv) Temperature drop at outside surface

$$= \frac{35}{4 \cdot 285} \times 0 \cdot 3 \qquad = 2 \cdot 17°$$

(b) *Homogeneous Wall* (Fig. 8.4)

$$R_T = Rsi + R + Rso$$

$$= 0 \cdot 7 + \frac{9}{8 \cdot 0} + 0 \cdot 3 \qquad = 2 \cdot 125 \text{ ft}^2 \text{ h } °F/Btu$$

$$td \qquad = 35°F$$

Gradient:

(i) Temperature at inner face of brick

$$= 65 - \left(\frac{35}{2 \cdot 125} \times 0 \cdot 7 \right)$$

$$= 65 - 11 \cdot 5 \qquad = 53 \cdot 5°F$$

(ii) Temperature at outer face of brick

$$= 53 \cdot 5 - \left(\frac{35}{2 \cdot 125} \times 1 \cdot 125\right) = 53 \cdot 5°F$$

$$= 53 \cdot 5 - 18 \cdot 5 \qquad\qquad = 35 \cdot 0°F$$

(iii) Temperature drop at outside surface

$$= \frac{35}{2 \cdot 125} \times 0 \cdot 3 \qquad\qquad = 5 \cdot 0°$$

The U values for these walls, in Imperial units, are:

Composite wall $\qquad \dfrac{1}{4 \cdot 825} = 0 \cdot 207 \text{ Btu/ft}^2 \text{ h } °F$

Homogeneous wall $\quad \dfrac{1}{2 \cdot 125} = 0 \cdot 47 \text{ Btu/ft}^2 \text{ h } °F$

Dewpoint gradient

As has been stated earlier, relative humidity is related to vapour pressures. If it is realized that atmospheric pressure is the sum of a number of partial pressures exerted by the various constituents of the atmosphere, it will be evident that water vapour, which is a gas, is itself a contributor to the whole. This partial pressure increases as the specific humidity increases, and the comparison between the *actual* vapour pressure and the *saturation* vapour pressure (i.e. the vapour pressure at saturation) for a given specific humidity gives the relative humidity, expressed as a percentage.

The *dewpoint* is the temperature to which a given atmosphere at a given specific humidity must be reduced to reach saturation: or, in other terms, it is the temperature at which the actual vapour pressure equals the saturated vapour pressure. Thus, if the gradients of actual vapour pressure and saturated vapour pressure through permeable materials (which includes most building materials) are drawn against the same scale, any point where the actual vapour pressure is higher than the saturated vapour pressure is liable to condensation. The calculation of these gradients is fairly complicated, though it follows much the same pattern as for temperature gradient, and we show below a calculation to produce an approximate dewpoint gradient which can be plotted against the temperature gradients described earlier. In this case, any point where the temperature falls below the dewpoint is liable to the risk of condensation.

Throughout the normal range of internal and external air temperatures (19° to −1°C (65° to 30°F)), the approximate dewpoint at the appropriate surface can be ascertained from the following table:

<table>
<tr><td colspan="3" align="center">Dewpoint =
Air Temperature (dry-bulb) minus 'a' when R.H. = 'b'</td></tr>
<tr><td colspan="2" align="center">a</td><td align="center">b</td></tr>
<tr><td align="center">°C</td><td align="center">°F</td><td align="center">%</td></tr>
<tr><td>0</td><td>0</td><td>100</td></tr>
<tr><td>1·4</td><td>2·5</td><td>90</td></tr>
<tr><td>3·0</td><td>5·4</td><td>80</td></tr>
<tr><td>5·1</td><td>9·2</td><td>70</td></tr>
<tr><td>7·3</td><td>13·1</td><td>60</td></tr>
<tr><td>9·7</td><td>17·4</td><td>50</td></tr>
<tr><td>12·9</td><td>23·2</td><td>40</td></tr>
</table>

FIG. 8.5. Approximate Dewpoints.

The dewpoints at the two surfaces can be established from this and the total dewpoint-difference across the construction ascertained. The total difference arises from the sum of the vapour resistances of the component materials.

The vapour resistivity of a material (the resistance to the passage of vapour per unit thickness) is the reciprocal of the vapour diffusivity, values of which for a number of common materials are listed in *Appendix* B. Thus, similarly to the method of thermal gradient calculation, the vapour resistivities multiplied by the thickness of material (producing the vapour resistance for the wall component) account proportionately for the total vapour resistance of the whole construction and, approximately in equal ratio, for the total dewpoint-drop.

Calculating the approximate dewpoint gradient for the wall shown in Fig. 8.3:

Composite Wall

Internal Air Temperature (d.b.)	$= 19°C$	$(65°F)$
Internal Relative Humidity	$= 50\%$	(50%)
Approx. Internal Dewpoint	$= 19° - 9·7°$	$(65° - 17·4°)$
(From Fig. 8.5.)		
	$= 9·3°C$	$(47·6°F)$
External Air Temperature (d.b.)	$= -1°C$	$(30°F)$
External Relative Humidity	$= 80\%$	(80%)
Approx. External Dewpoint	$= -4·2°C$	$(26·2°F)$
Dewpoint difference (surface-to-surface)	$= 13·5°$	$(21·4°)$

Vapour Diffusivities (d):

	gm.mm/m².s.bar	lb in/ft²h.atm
Fibreboard	1·65	0·05
Brickwork	4·05	0·12

Therefore Vapour Resistivities $\left(\dfrac{1}{d} = vr\right)$:

	m².s.bar/gm.mm	ft² h.atm/lb in
Fibreboard	0·60	20·0
Brickwork	0·25	8·3

And Vapour Resistances (L × vr = vR):

	m².s.bar/gm	ft² h.atm/lb
Fibreboard	$25 \times 0·60 = 15·0$	$1 \times 20 = 20·0$
Brickwork	$225 \times 0·25 = 56·25$	$9 \times 8·3 = 74·7$
Total vR	71·25	94·7

From this, the approximate dewpoint at the interface between the fibreboard and the brick can be calculated and the approximate gradient drawn:

$$\text{Dewpoint-drop across brickwork} = \frac{56·25}{71·25} \times 13·5° \qquad \left(\frac{74·7}{94·7} \times 21·4°\right)$$

$$= 10·6° \qquad (16·6°)$$

$$\text{Dewpoint at interface} = -4·2° + 10·6° \qquad (26·2° + 16·6°)$$

$$= 6·4°C \qquad (42·8°F)$$

The approximate dewpoint gradient is plotted against the

temperature gradient in Fig. 8.6, and it will be seen that though the dewpoint at the internal surface is well below the surface temperature, the dewpoint at the interface is $1.28°C$ ($2.0°F$) above temperature at the same point. The hatched area where dewpoint is above temperature is the area in which condensation is likely to occur. It will be appreciated that, if different relative humidities had been adopted, a different gradient would have resulted and consequently a different area of condensation risk area would have been found.

In the same way that standard assumed values for external and internal temperatures are taken for thermal calculations, standard humidities can be assumed for vapour calculations. Internal humidities in buildings which are not air-conditioned vary with the weather but, since condensation problems usually arise in cold weather, rather than warm, it is reasonable to take external relative humidities of 80 to 100 per cent (fog conditions). Domestic internal relative humidities generally range from 40 to 60 per cent at $19°C$ ($65°F$). Offices, schools and similar uses are in the same range, as are light industrial buildings, but certain processes create high humidities (e.g. paper mills, laundries and dairies), while others require high humidities for technical reasons connected with the process (e.g. textile mills). For these buildings an internal relative humidity of 75 per cent at $19°C$ ($65°F$) would be reasonable, and if a surface temperature drops below $14°C$ ($57°F$) condensation on the surface will result.

Another factor, which has an important bearing on the determination of the risk of condensation, is the reduction of thermal resistivity of permeable materials when wet. Thus, if a brick wall is wetted by rain its resistance falls and consequently the internal surface temperature also falls. Values of 'k' and 'r' are usually calculated for unwetted conditions and it is therefore more practical, when calculating for condensation risks, either to underestimate thermal resistance or to ensure that the temperature gradient keeps comfortably above the dewpoint gradient.

Vapour barriers

The insertion of a vapour barrier in a composite wall is designed to drop the dewpoint gradient below the temperature gradient. Fig. 8.7 shows a wall similar to that in Fig. 8.6, except that two

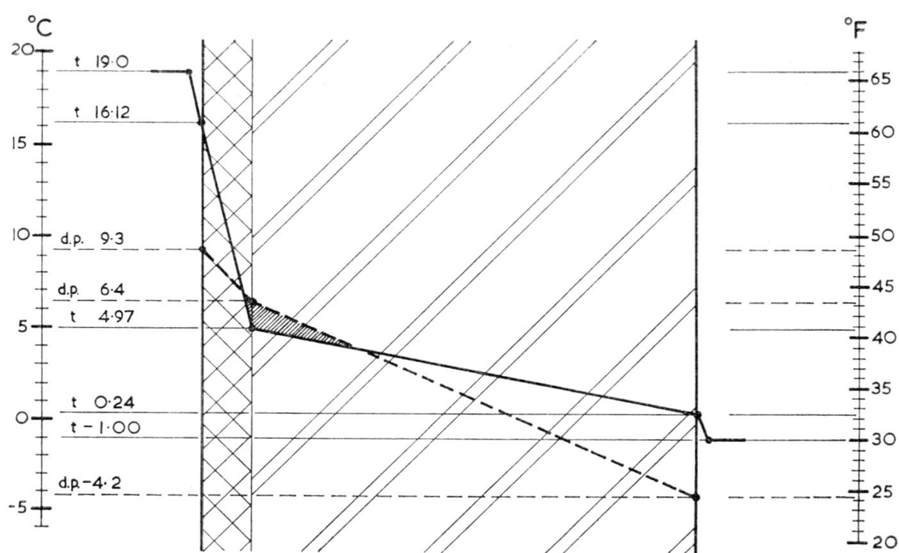

FIG. 8.6.   Ascertainment of Condensation Risk.

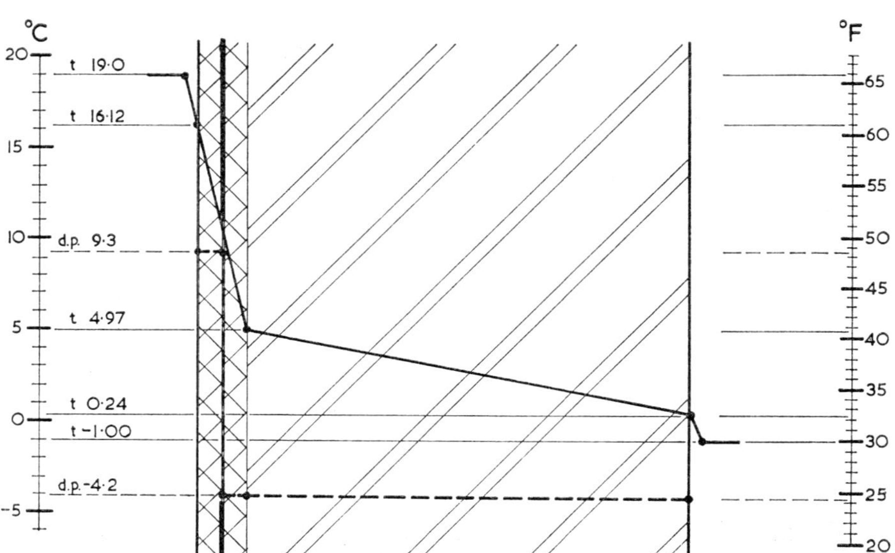

FIG. 8.7.   Effect of Foil Vapour Barrier.

12 mm insulation boards with an aluminium foil layer between are used in place of one 25 mm board.

The temperature gradient remains substantially as before, but the dewpoint gradient must be recalculated as follows:

Approx. Internal Dewpoint $=$ 9·3°C   (47·6°F)
Approx. External Dewpoint $=-4·2$°C   (26·2°F)
Dewpoint difference (surface-
   to-surface) $=$ 13·5°   (21·4°)

Vapour Diffusivities (d) and Resistivities (vr) for insulating board and brickwork remain as before, but the Vapour Diffusance of aluminium foil is 0·00003 gm/m².s.b (0·00002 lb/ft² h atm) and its Vapour Resistance is therefore 33,000 m².s.b/gm (50,000 ft² h atm/lb). The total Vapour Resistance across the wall is therefore:

12 mm fibreboard	$12 \times 0·6 = 7·2$	$(\tfrac{1}{2} \times 20 = 10·0)$
Aluminium foil	33,000	$($ 50,000$)$
12 mm fibreboard	$12 \times 0·6 = 7·2$	$(\tfrac{1}{2} \times 20 = 10·0)$
Brickwork	$225 \times 0·25 = 56·25$	$(9 \times 8·3 = 74·7)$
Total vR	33,070·65	50,094·7

Dewpoint-drop across brickwork

$$\frac{56·25}{33,070·65} \times 13·5 \qquad \left(\frac{74·7}{50,094·7} \times 21·4\right)$$

$$= 0·023° \qquad\qquad (0·032°)$$

Dewpoint at interface brick/board

$$-4·2 + 0·023 \qquad (26·2 + 0·032)$$
$$= -4·177°C \qquad (26·232°F)$$

Dewpoint-drop across 12 mm board

$$\frac{7·2}{33,070·65} \times 13·5 \qquad \left(\frac{10}{50,094·7} \times 21·4\right)$$

$$= 0·003° \qquad\qquad (0·0045°)$$

Dewpoint at interface board/foil

$$-4·177 + 0·003 \qquad (26·232 + 0·0045)$$
$$= -4·174°C \qquad (26·2365°F)$$

Dewpoint-drop across aluminium foil

$$\frac{33,000}{33,070·65} \times 13·5 \qquad \left(\frac{50,000}{50,094·7} \times 21·4\right)$$

$$= 13·471° \qquad\qquad (21·359°)$$

Dewpoint at interface foil/board

$$- 4{\cdot}174 + 13{\cdot}471 \qquad (26{\cdot}2363 + 21{\cdot}3590)$$
$$= 9{\cdot}297°C \qquad\qquad (47{\cdot}5955°F)$$

Dewpoint-drop across 12 mm fibreboard

$$\frac{7{\cdot}2}{33{,}070{\cdot}65} \times 13{\cdot}5 \qquad \left(\frac{10}{50{,}094{\cdot}7} \times 21{\cdot}4\right)$$
$$= 0{\cdot}003° \qquad\qquad (0{\cdot}0045°)$$

Dewpoint at inner-wall surface

$$9{\cdot}297 + 0{\cdot}003 \qquad (47{\cdot}5955 + {\cdot}0045)$$
$$= 9{\cdot}3°C \qquad\qquad (47{\cdot}6°F)$$

It will be appreciated that the minute drops in dewpoint across the brickwork and insulation board are not significant, and that the tremendous vapour resistance of the aluminium foil accounts for virtually all of the drop. Hence, when a vapour barrier of such extreme performance is used, the dewpoint line may be drawn as horizontal through all the remaining materials. Fig. 8.7 shows this sharp fall in the dewpoint line and demonstrates how the condensation risk apparent in Fig. 8.6 has been eliminated, though a small change in position could re-introduce the risk.

Less effective vapour barriers than metallic foils will also cause sharp drops in the dewpoint gradient, and Fig. 8.8 shows a wall where the hidden layer of fibreboard of the previous example is replaced by a 12 mm layer of expanded polystyrene, and the inner layer by 12 mm plasterboard.

In this case, the vapour barrier has a thermal value, too, and, therefore, as the dewpoint gradient drops sharply, so does the temperature gradient. The positioning of the barrier now becomes of less importance, and Fig. 8.9 shows the barrier in a different position, where the risk of condensation is still avoided because the two gradients lie approximately parallel through all materials.

Thus, the positioning of a partial vapour barrier is usually less critical than that of a highly efficient vapour barrier whose thermal conductivity is high. The efficient vapour barrier must be kept as close as possible to the warm side of the wall.

Partial vapour barriers whose thermal conductivity is low may still produce condensation risks if their vapour diffusivity results in a dewpoint gradient shallower than the temperature

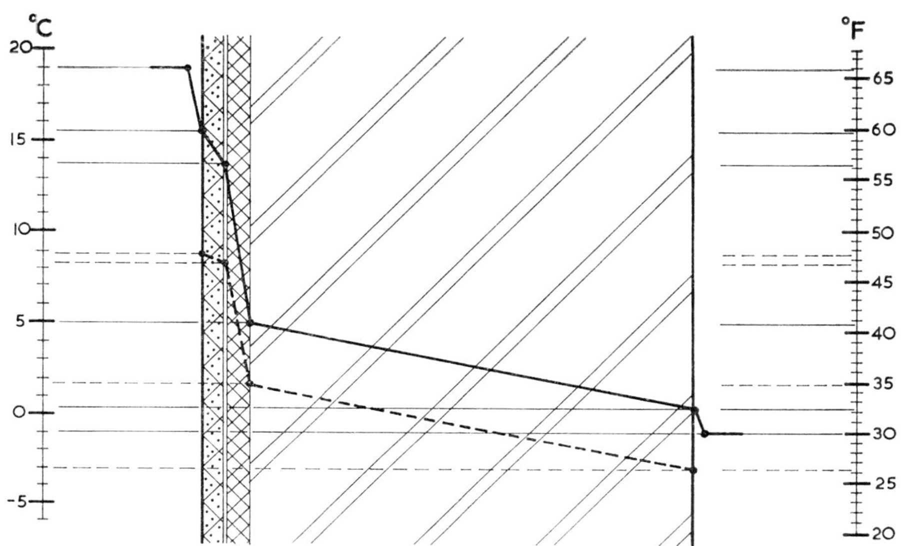

FIG. 8.8.   Effect of Partial Vapour Barrier on Dewpoint Gradient.

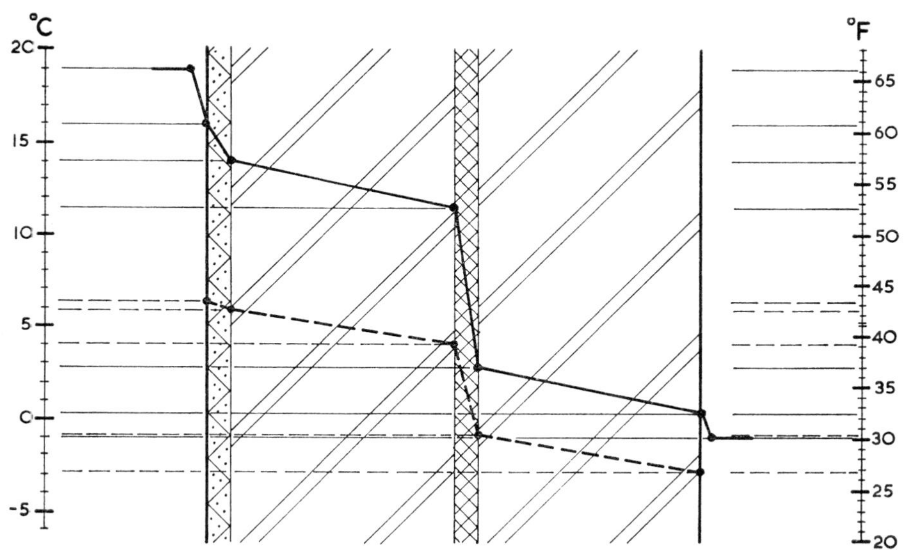

FIG. 8.9.   Effect of Partial Vapour Barrier on the Dewpoint Gradient
with Barrier in different position.

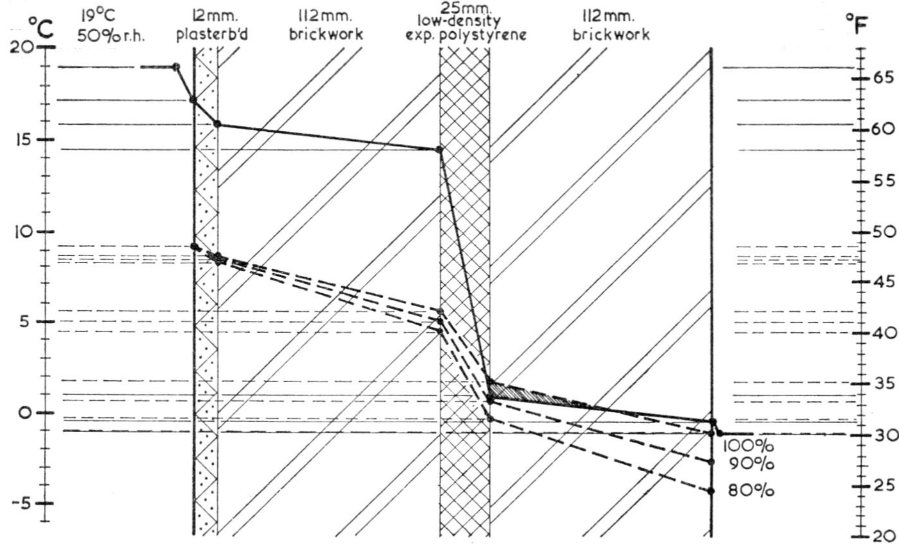

Fig. 8.10.   Effect of External Relative Humidity on Condensation Risk.

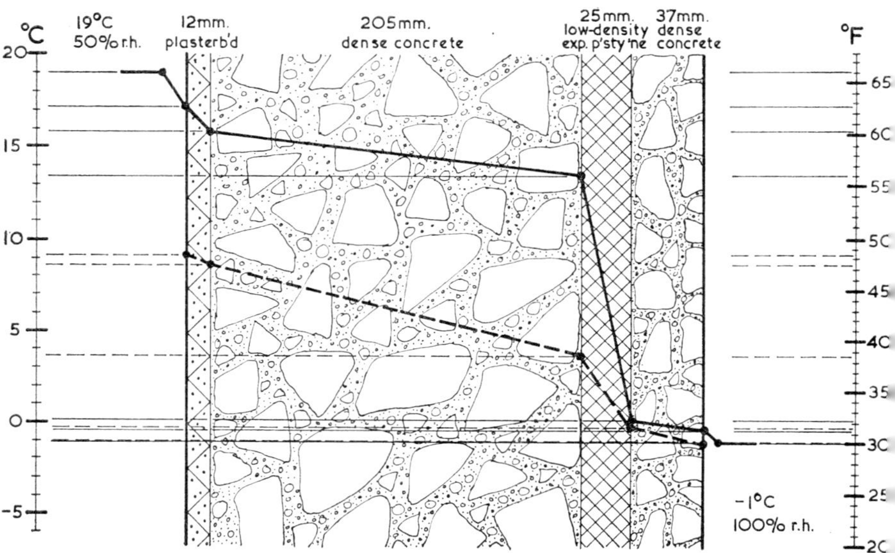

Fig. 8.11.   Effect of moving Partial Vapour Barrier towards cold side of Wall.

gradient. Fig. 8.10 shows a similar wall to that in Fig. 8.9, but where the vapour barrier is of 25 low-density expanded polystyrene, whose vapour resistance is relatively low and whose thermal resistance is high. The gradients for 80, 90 and 100 per cent external relative humidity are plotted and it will be seen that, when external conditions are very damp, the risk of condensation appears.

The general rules, then, for the positioning of vapour barriers may be stated as follows:

1. *Where the vapour barrier has high vapour resistance and low thermal resistance, it must be placed close to the warm side of the structure.*
2. *Where the vapour barrier has high vapour resistance and high thermal resistance, it may be placed at any point in the wall.*
3. *Where the vapour barrier has moderate vapour resistance and high thermal resistance, it must be placed near the cold side of the wall.* (*See* Fig. 8.11.)

Other constructions

So far we have considered walls of fairly conventional construction, but problems of condensation can arise in all forms of construction. Fig. 8.12 shows a roof comprising open-web steel beams supporting a metal-sheet roof-deck with fibreboard and asphalt. The ceiling shown is foil-backed fibreboard.

If the ceiling were of ordinary fibreboard, without the vapour barrier, the dewpoint would be higher through the structure and a risk of condensation would exist in the cavity – primarily on the underside of the decking, which, though a vapour barrier in itself, could be subject to low air temperature but relatively high dewpoint. The presence of the foil-vapour barrier immediately above the ceiling-board reduces the dewpoint in the cavity and thus eliminates the risk.

It must be noted, however, that it is possible for the temperature of the exposed surface of a roof to fall below external air temperature, particularly on clear nights when radiant-heat loss is not counterbalanced by radiant-heat gain from clouds. In these circumstances condensation can form above the metal deck (or concrete roof) causing blistering or fracture of the felt or asphalt roof. Various methods of ventilation or drainage of this moisture are available.

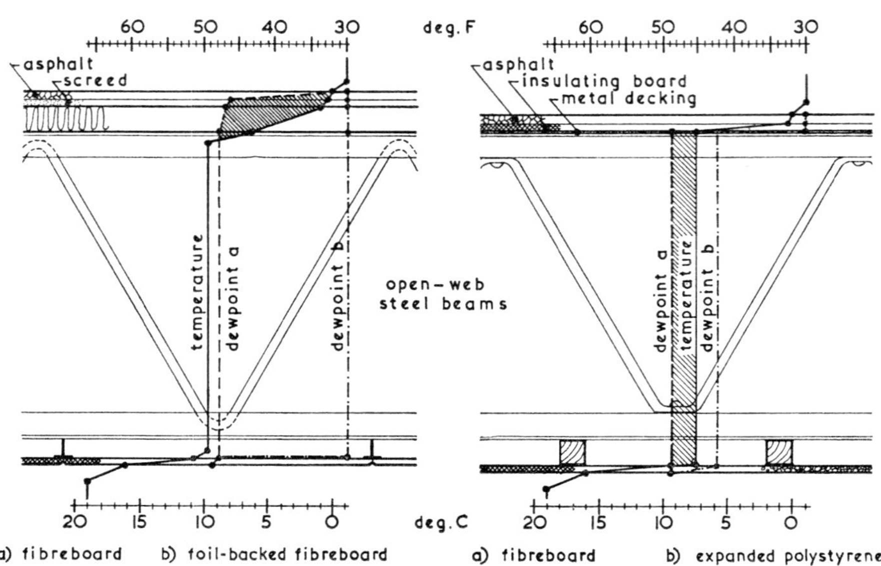

FIG. 8.12.   Vapour Barrier in Hollow Roof Structures.

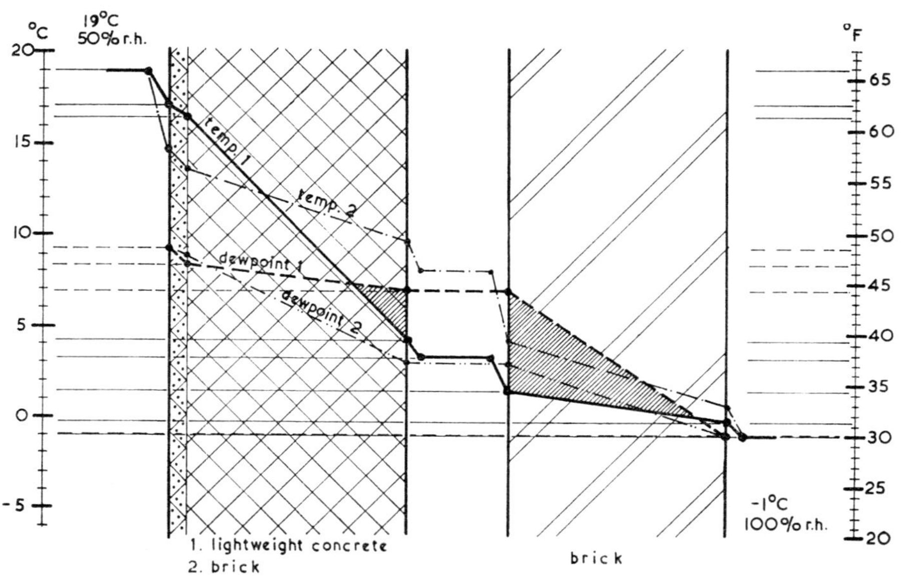

FIG. 8.13. Condensation Risks in Cavity Walls.

Cavities

The effect of cavities on dewpoint has not been discussed so far, but the foregoing example provides a clue to their effect on the temperature–dewpoint gradients. The diffusivity of air is high (18·0 gm mm/m² s b or 3·7 gr in/ft² h mb), since the equalization of vapour pressures meets little resistance; thus, while the two thermal surface-resistances add to the thermal efficiency of the construction, the rapid equalization of vapour pressures across the cavity maintain an almost flat dewpoint gradient. Fig. 8.13 shows a cavity wall, consisting of a lightweight concrete inner leaf and brick outer leaf, with appropriate temperature and dewpoint gradients plotted. The risk of condensation on both sides of the cavity is apparent. Normally this condensation, if heavy, runs down the wall to the weathered top surface of the cavity filling and is discharged through the weep holes. It should be noted, however, that as the condensation develops some of it is absorbed by the brickwork, thus reducing the thermal resistance, dropping the thermal gradient still further and aggravating the condensation risk. Thus, 'steady-state' conditions for such a wall have to allow for a lower U value when external relative humidity is high.

The replacement of the inner leaf by brickwork alters the relationship of the two gradients so that the dewpoint remains below the temperature (as shown by the two alternative gradients 'Temp 2' and 'd.p.2', in Fig. 8.13).

This underlines the rules laid down above, where Rule 3 states that a vapour barrier of moderate vapour resistance and high thermal resistance must be placed near the cold side of the wall. The lightweight concrete has only moderate vapour resistance and, though its thermal resistance is also moderate, the combination of the inner leaf with the cavity produces an overall result similar to that produced by a partial vapour barrier of high thermal resistance.

Condensation on glass

Windows are usually the first surfaces to suffer from condensation, and wood windows may be damaged by wet-rot, steel windows by rust, as a result. Both types may suffer in appearance by damage to paintwork and unsightly pools of water on sills. The early onset of condensation with thin impermeable membranes of low thermal resistance follows from the calculations made above concerning walls and roofs. Typical calculations follow:

Single Glazing

k　　　= 1·05 Wm/m² deg C　　　(7.3 Btu in/ft² hr deg F)
Glass = 7·3 kg/m²　　　　　　　(24 oz/ft²)
　　　@ 2·5 mm thick　　　　　　($\frac{1}{10}$ in)

Internal temperature　19°C (65°F) Internal R.H.　50%
　　　　　　　　　　　　　　　　Dewpoint　9·3 (47·6)

External temperature　−1°C (30°F) External R.H.　80%
　　　　　　　　　　　　　　　　Dewpoint −4·0 (26·2)

Temp. difference　　　20°C (35°F)

$$R_T \ = 0·121 + \frac{0·0025}{1·05} + 0·052 \ = 0·17538$$

$$(0·7 + 0·0137 + 0·3 \ = 1·0137)$$

$$Tsi \ = 19 - \left(\frac{20}{0·17538} \times 0·121\right) \ = 5·18°$$

$$(65 - 24·17 = 40·83°)$$

Thus, the surface temperature of the glass is 4·12° below dewpoint and condensation will occur.

Double Glazing

Two sheets of glass, as above, with sealed air-gap between. Temperatures and relative humidities as before:

$$R_T \ = 0·121 + \frac{0·0025}{1·05} + 0·173 + \frac{0·0025}{1·05} + 0·052$$

$$= 0·351 \ (2·027)$$

$$Tsi \ = 19 - \left(\frac{20}{0·351} \times 0·121\right) \ = 12·09° \ (52·92)$$

Thus, the surface temperature of the inner glass is nearly 3° (5·5°) above dewpoint and condensation will not occur.

Note that the vapour diffusivity of glass is extremely low and therefore the dewpoint gradient falls very sharply, as it does when an efficient vapour barrier is used. A dewpoint gradient need not be drawn to establish surface condensation, though, with double glazing, the possibility of condensation within the sealed cavity must be considered. Most double-glazed units use aluminium vapour seals at the edges to prevent humidification of the dry air sealed in the cavity at manufacture; otherwise the cavity air would, over a period of time, reach a humidity where the dewpoint at the inner surface of the outer sheet of glass could be above air temperature, with resultant inaccessible condensation.

9: Thermal Characteristics of Buildings

IN THE TWO FOREGOING CHAPTERS, we have dealt with the considerations to be evaluated in selecting constructions which will resist the transmission of heat and vapour, and we have shown that the type and position of the thermal insulant and vapour barrier have an effect on surface temperature and moisture concentration. A further factor in the selection of the materials to be used and their relationship with the inner and outer surfaces is the thermal inertia of the structure.

It should now be fully appreciated that, depending upon the resistance of the structure to heat flow and the amount of mass within any insulant layer, a building will respond either slowly or quickly to changes in conditions within, or outside, the structure. Internal changes may be caused by artificially imposed heat gains or the cessation of existing imposed gains, while external changes result from weather variations.

Fig. 9.1 shows the general situation for a wall in both summer and winter conditions. The manner of heat flow involves radiation, conduction and convection, but these can all be represented by the arrow of net heat flow.

The transmission of heat by conduction within a material occurs, as has been stated earlier, by the transfer of energy from one particle to another, the quantity of energy per transfer being a part of the difference between the energy states of the two particles. Thus, each particle retains some of the energy it receives and passes on the remainder. As a consequence, throughout the section of a material subjected to heating from one side, there is a gradient of energy levels. This gradient, when compared with a lower temperature level, represents a quantity of heat stored in the material. Fig. 9.2 illustrates this by showing the difference between the energy levels at midday and at midnight. The hatched area represents a quantity of heat energy

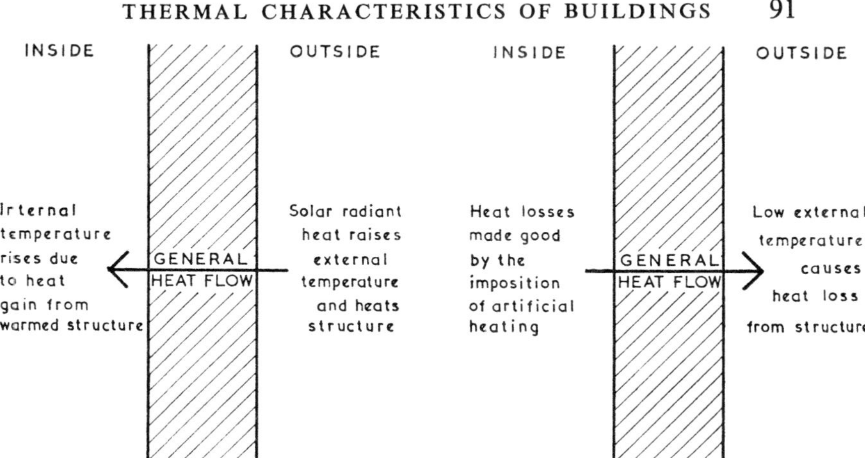

FIG. 9.1.   Seasonal Heat Flow.

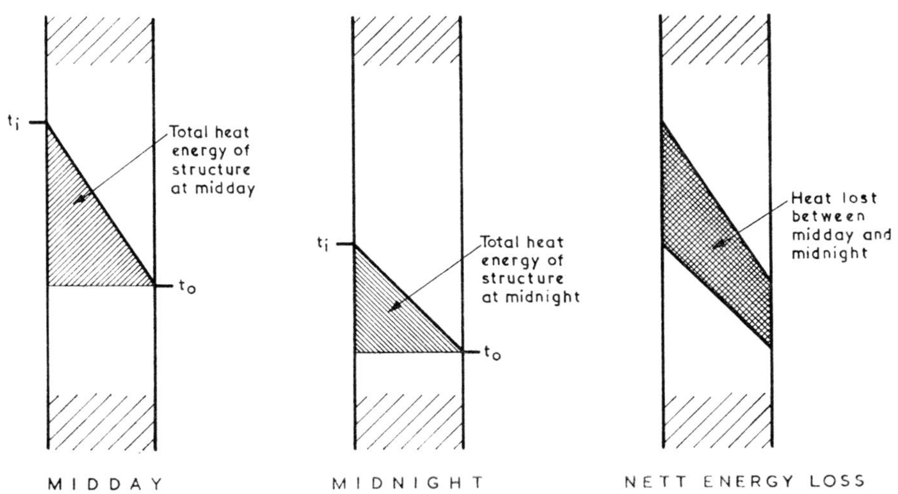

FIG. 9.2.   Diurnal Variation in Energy Gradients.

(N.B. Surface effects are disregarded.)

which, if gained during the day, will be lost as the temperature drops.

Thus, the transmittance of heat energy through a structure results in a build-up of energy in the structure. Similarly, solar radiant heat gains during a summer day create a reservoir of heat energy in the wall which dissipates when the heating ceases to be applied. Dependent upon the total quantity of heat stored in the structure, and upon the rate at which it can release the heat (i.e. its conductivity), the total quantity takes a certain time to be released. Equally, it takes a certain time to be stored up.

Radiant heat passing through glazing in a building impinges on interior structure, and heat artificially imposed imparts energy to both external and internal structures. So it will be seen that all elements of the building can contain these stores of energy, and it follows that the release of heat energy, after a change of condition, will occur from the whole mass of the structure.

The intensity of this release of energy will depend upon the nature of the material and its conductivity; the actual quantity of heat emitted will depend upon its thermal storage characteristics.

Calculation of the time constant

It will be inferred, from the foregoing, that each building has a thermal response related to the thermal characteristics of that particular building and its use. The *I.H.V.E. Guide to Current Practice*, in the section on Meterological Data, attempts to correlate all the variables associated with this problem. However, it is sufficient at this stage to appreciate that a building responds in this way and that the heat stored within the structure will take some time to dissipate after the heating system has been shut down. This is referred to as the *time constant* for a building and is given by the ratio:

$$\frac{\text{Heat stored in the structure,}}{\text{Hourly Heat Loss from it}}$$

expressed in terms of degree difference between the inside and the outside temperature (i.e. between the internal and the external climate).

The example given on the two following pages will serve to indicate the procedure adopted:

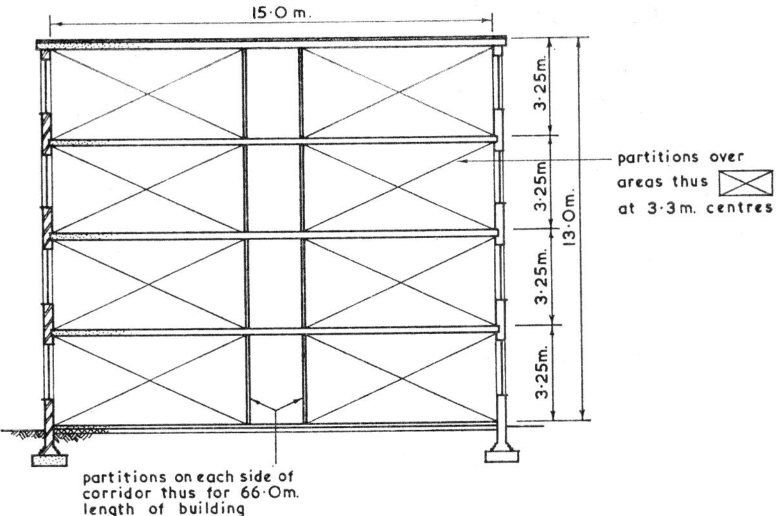

FIG. 9.3.   Example of Construction for Calculation of Time
Constant (in S.1. Units.)

Details of Building

Element	Construction	U Value
Walls	340 mm brick with 15 mm plaster inside	1·76
Roof	Asphalt 150 mm concrete with 12 mm cork lining	1·76
Windows	Single glass 3 mm	5·68
Ground Floor	150 mm concrete on hard core	1·14
Intermediate Floors	200 mm solid concrete	—
Partitions	75 mm hollow clay blocks	—

Calculation

With ventilation taken at two air changes per hour, and
window area at 40% of wall area, the calculation will be as
follows:

Heat Loss per deg C difference

kW

$$\text{Air} \quad \frac{2}{3600} \times 13 \times 15 \times 66 \times 1200 \qquad = 8 \cdot 580$$

Walls $0 \cdot 6 \times 2 \times 13 \times (66 + 15) \times 1 \cdot 76 \quad = 2 \cdot 224$
Glass $0 \cdot 4 \times 2 \times 13 \times (66 + 15) \times 5 \cdot 68 \quad = 4 \cdot 785$
Roof $66 \times 15 \times 1 \cdot 76 \qquad\qquad\qquad = 1 \cdot 742$
Floor $66 \times 15 \times 1 \cdot 4 \qquad\qquad\qquad\; = \underline{1 \cdot 129}$

$$= \underline{\underline{18 \cdot 460}} \qquad (i)$$

Heat Stored in Building

$$t_i \quad = 19°C$$
$$t_o \quad = -1°C$$
$$t\text{ ground} = 10°C$$

Element	Mean Temperature °C	Weight kg	Specific Heat J/kgdegC	Density kg/m³	Heat Content kJ
Roof: Cork ..	9·7	1,710	1·80	144	32,934
Concrete	3·0	396,400	0·88	2400	1,395,328
Asphalt	0·6	39,917	0·84	2240	53,648
Walls	7·4	790,240	1·05	1760	6,969,917
Glass	4·7	6,300	0·84	2520	30,164
Ground Floor	13·8	396,400	0·88	2400	1,325,561
					9,807,552
Intermediate Floors ..	19·0	1,425,000	0·88	2400	25,080,000
Partitions ..	19·0	330,109	1·05	67 kg/m²	6,932,289
Total heat stored in building for 20° difference					41,819,841

$\therefore$ Heat stored in building for 1°C difference between inside and outside $= 2,090,992$ kJ .... (ii)

$$\text{Time constant for building} = \frac{(ii)}{(i) \times 3600} = \frac{2,090,992}{18 \cdot 46 \times 3600}$$

$$= 31 \cdot 4 \text{ hours}$$

The following parallel calculation in Imperial units, taken from a paper by H. C. Jamieson in the *Journal* of the I.H.V.E., is for a similar building:

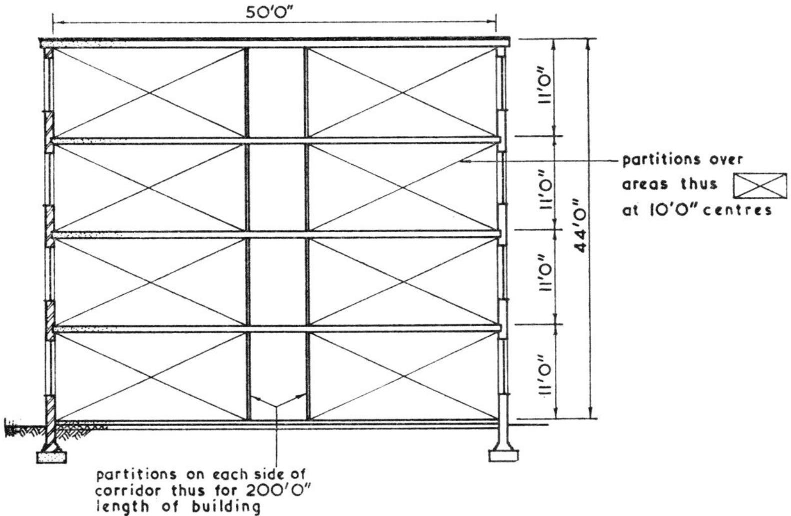

FIG. 9.4.   Example of Construction for calculation
of Time Constant (in Imperial Units)

Details of Building

Element	Construction	U Value
Walls	$13\frac{1}{2}$ in brick plastered inside	0·31
Roof	Asphalt. 6 in concrete. $\frac{1}{2}$ in cork	0·31
Windows	Single glass $\frac{1}{8}$ in	1·00
Ground Floor ..	6 in concrete on hard core	0·20
Intermediate Floors	8 in solid concrete	—
Partitions	3 in hollow clay blocks	—

Calculation

With air changes taken at two per hour, and glass area taken at 40 per cent of wall area, the calculation will be as follows:

Heat Loss per deg F difference

		Btu/hr
Air	$2 \times 44 \times 50 \times 200 \times 0.019$	16,700
Walls	$0.6 \times 2 \times 44 (200 + 50) \times 0.31$	4,100
Glass	$0.4 \times 2 \times 44 (200 + 50) \times 1$	8,800
Roof	$200 \times 50 \times 0.31$	3,100
Floor	$200 \times 50 \times 0.2$	2,000
	Total heat loss per deg F difference	34,700 ... (1)

Heat Stored in Building

Internal temperature taken at 65°F.

Datum temperature taken at 30°F except for floor, where it is taken at 50°F.

Heat stored in internal beams and stanchions is ignored.

Element	Mean Temperature °F	Weight lb	Specific Heat	Density lb/cu. ft	Heat Content Btu
Roof: Cork ..	49·5	3,750	0·43	9	31,400
Concrete	37·2	750,000	0·21	150	1,140,000
Asphalt	33·4	87,500	0·2	140	59,500
Walls	45·4	1,635,000	0·25	110	6,300,000
Glass	40·5	14,400	0·2	157	30,300
Ground Floor	56·9	750,000	0·21	150	1,085,000
					8,646,200
Intermediate Floors ..	65·0	3,000,000	0·21	150	22,050,000
Partitions ..	65·0	702,000	0·25	124 lb/ sq yd	6,120,000

Heat stored in building for temperature difference between inside and outside of 35°F:

36,816,200 Btu

Heat stored in building for 1°F difference between inside and outside:

1,050,000 Btu ....(2)

$$\text{Time-constant for building} = \frac{(2)}{(1)} = \frac{1,050,000 \text{ Btu}}{34,700 \text{ Btu/hr}}$$

$$= 30.4 \text{ hours}$$

Classification of Buildings

According to the time resulting from such calculations, a building is classified A or B in the I.H.V.E. Guide.

This affects the percentage overload required for the heating system and the outside temperature to be adopted. In the case of the example just given, the Thermal Time Lag would approach one and a half days. This would be the theoretical time taken for the building to give up all its heat and, by reference to the current *I.H.V.E. Guide*, the designer has a basis for starting his designs.

Classification

Type of Building	Classification
Multi-storey buildings with solid floors and internal partitions	A
Single Storied Buildings	B

U.K. Winter External Design-Temperature Determination

Building Thermal Inertia Classification	System Overload Capacity %	External Design Temperature °C	°F
A	20	−1	30
B	20	−3	27
A	nil	−4	25
B	nil	−5	23

If this reasoning applies to a structure during the winter when artificial heat is supplied from within the building, it must also apply to the structure during the summer, except in as much as the heat will be applied externally.

The important thing is that these heat gains during the summer are transient, altering throughout the day with maximum to minimum heat gains related to the sun's position.

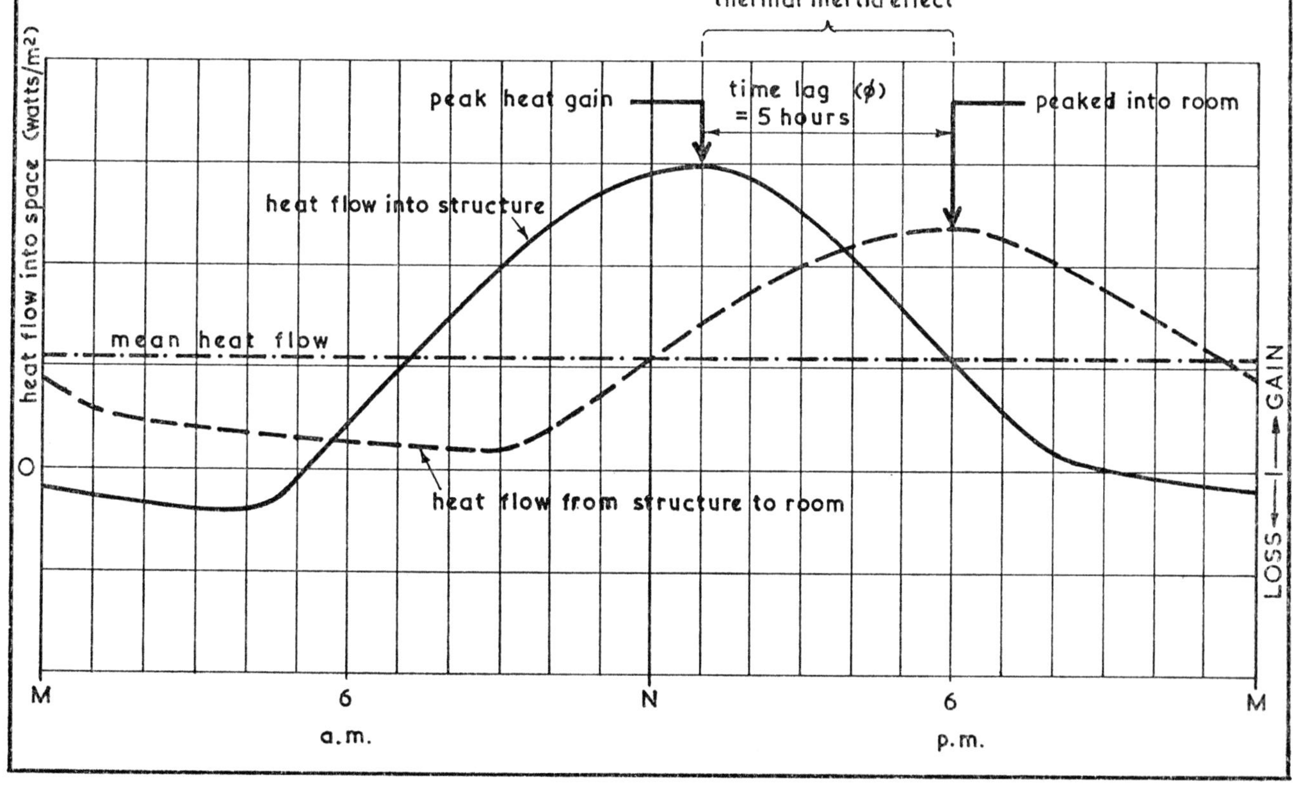

Fig. 9.5.   Graph of Heat Flow through Horizontal Roof.

Also, the solar heat gains have two components, namely:

(i) Direct radiation – sun on structure
(ii) Indirect radiation:
    (*a*) Sky – re-radiated heat
    (*b*) Ground – re-radiated heat.

The effect of these will vary with the type of external cladding. For example: glass (whether single-glazed or double-glazed) will permit instantaneous heat gains to occur within a space from both elements of radiation, while dense concrete construction with insulating layer will only transmit the effects of the solar gain after a period of time.

This is indicated by reference to the graph opposite (Fig. 9.5) showing the effects of time lag on a 150 mm thick roof slab, and the 'effective' time at which the maximum heat gains could be expected to influence the internal room condition. The time lag Φ was derived from the density of slab and reference to the chart (Fig. 9.6).

Figure 9.7 (page 100) shows variations in temperature and humidity during a typical summer week at the end of August.

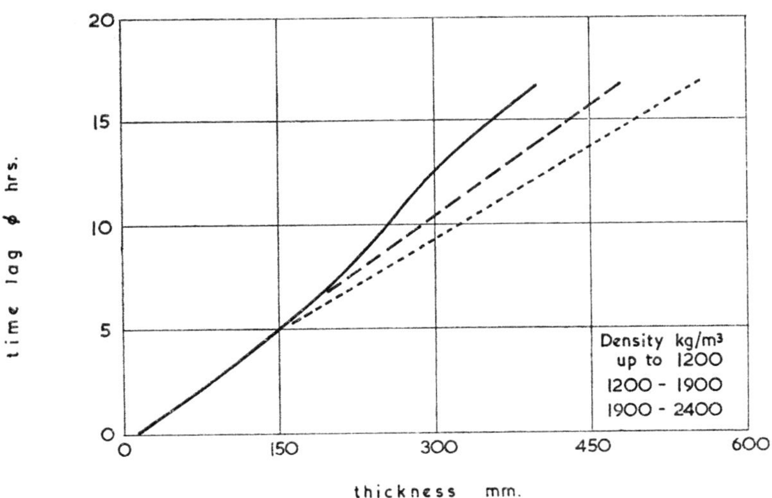

FIG. 9.6.   Chart of Variation of Time Lag with Thickness.

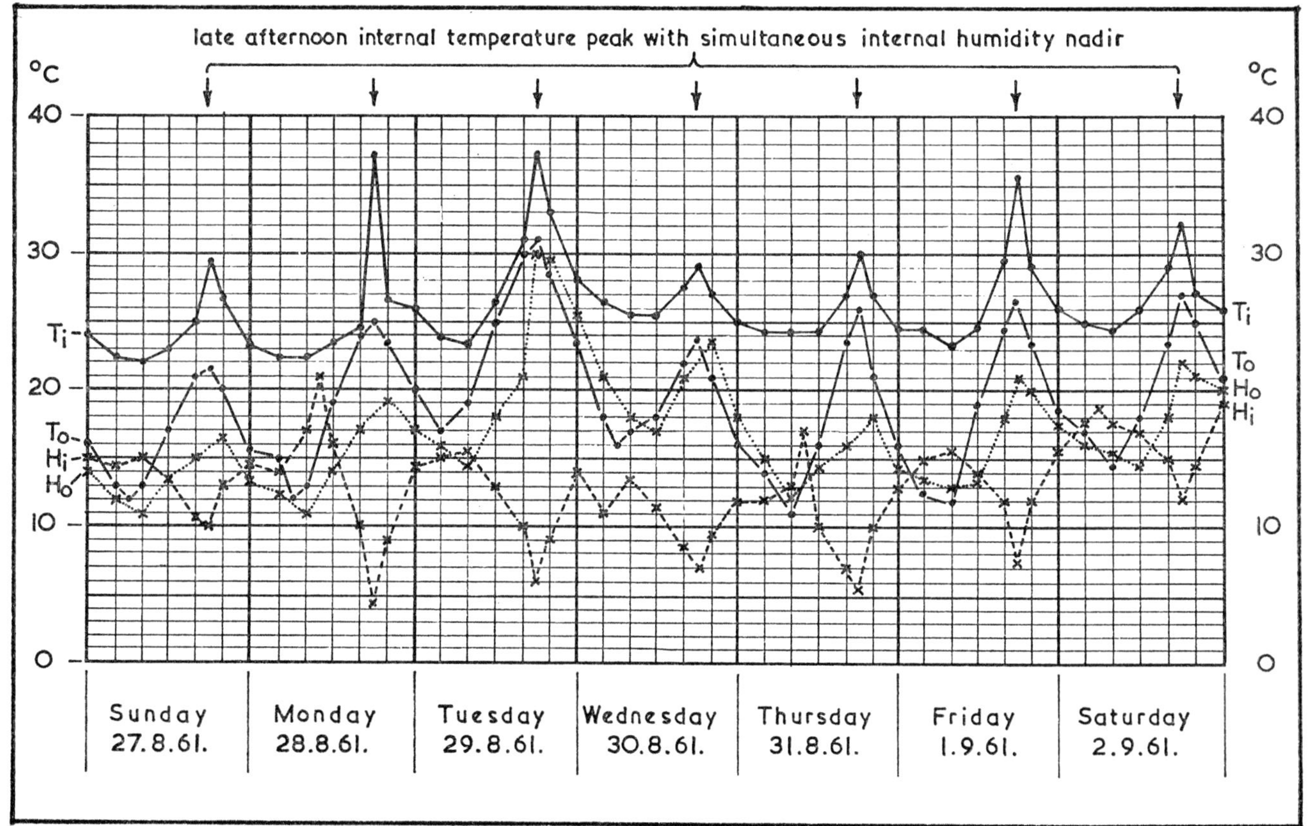

Fig. 9.7. Diurnal Variation in Temperature and Humidity.

10: The Selection of Suitable Design Bases

IN CHAPTER 7, A NUMBER OF design indices were listed and described as to their general applicability to the circumstances and climate. It will be evident that these indices can only generalize about the requirements for comfort, and that different groups of people have different standards in the same way as individuals differ in their metabolism. The parameters which must be used, therefore, should be selected in accordance with the ascertained or presumed requirements of the particular group concerned. Thus, a number of fundamental considerations must be reviewed to produce a general basis, followed by detailed consideration of the particular problem.

Fundamental considerations

1. Statutory requirements:
   (i) Acts of Parliament.
      (*a*) *Thermal Insulation (Industrial Buildings) Act, 1957.* This sets levels of temperature and thermal insulation for industrial building.
      (*b*) *Offices, Shops and Railway Premises Act, 1963.* This sets levels of temperature in such buildings as well as space requirements for occupants of offices.
      (*c*) *Public Health Acts, 1936 and 1961.*
      (*d*) *Public Health (Scotland) Acts, 1945 and 1962.* Both Acts contain enabling clauses under which the Building Regulations are made.
      (*e*) *London Building Acts, 1930, 1935 and 1939.* These Acts contain enabling clauses under which the G.L.C. Bye-laws are issued. The Acts contain some controls over the volume and heights of habitable rooms, and some more recent regulations regarding car parking buildings include controls of ventilation.

(ii) Bye-laws. *The London Building (Constructional) Bye-laws, 1964*. Section 13.05 requires ventilation openings in certain plant rooms, but this is connected with fire precautions rather than environment.

(iii) Building Regulations. Both the Scottish Regulations and those for England and Wales include comprehensive sections on thermal insulation and space and ventilation.

(iv) Codes of Practice. *B.S.C.P.* 3 (Chapter VIII) contains heating design considerations, recommended comfort standards and insulation standards as well as relating these standards to the economics of the building.

2. Obligatory standards set by authorities bodies: e.g. Ministry of Health – Hospital Building Notes; Department of Education – Design Bulletins.

3. Recommended standards advised by professional bodies: e.g. *I.H.V.E. Guide* (U.K.); *A.S.H.V.E.* (U.S.A.).

Detailed considerations

1. Type of Occupation:
    (i) Sleeping (hospitals, hostels)
    (ii) Sedentary work (offices, light assembly factories, schools)
    (iii) Light work (factory machine-shops, school workshops)
    (iv) Heavy work (steel fabrication, warehousing, etc.).

2. Special Applications:
    (i) Operating theatres
    (ii) Computers
    (iii) Laboratories
    (iv) Process work
        . . . and many others.

In many cases the specific requirements of the building may be known to the client (particularly where the client is a public authority) or his special advisers. The lay client often relies upon the architect, together with the engineer, to build up a brief for acceptance by him. It is therefore necessary to follow a logical path from the fundamental to the particular and to apply a sound basic knowledge of thermal environmental requirements to produce a rational proposal.

Reference to the appropriate statutory regulations is, of

course, the first step. Next, enquiring of the appropriate Ministry as to design requirements for problems within their scope, and then reference to professional text-books.

The detailed considerations may often be ascertained from the investigation of fundamentals, as with hospitals and schools, but in industry and commerce less complete information is available and the brief has to be constructed by analysis of the functions involved.

Occupation

(i) *Sleeping*. In hospitals the patients must be regarded as immobile and it is therefore vital that the control of heating ventilation, solar gain and glare should be considered thoroughly. It might be that the provision of permanent sun-breakers to ward windows would be the only satisfactory control of the last two aspects, whereas in a hostel building the occupant can pull the curtains or drop the blind for himself. In any event, the metabolism of patients is often at a low ebb and the heating system must be adequate to maintain a higher temperature than for other buildings, if comfort is to be achieved. At the same time, the nurses have a busy time and require a brisk atmosphere to prevent lassitude; so the ventilation rate must be adequate.

(ii) *Sedentary work*. Even sedentary work involves a certain amount of physical effort and the rate of bodily heat-production is greater than for sleeping. Thus a lower temperature is required, though the ventilation rate must be sustained if efficiencies are to be maintained.

(iii) *Light work*. Much depends upon the intensity of physical activity and some judgement is needed to arrive at suitable levels. For example, while school classrooms may be regarded as sedentary, the workshops and gymnasia are not. The workshops may require industrial standards of environment with dust-arrestor plant in the joinery area, while the gymnasia may need a moderately high temperature for eurythmics and a lower temperature for games; but in both cases they will require a high standard of ventilation.

(iv) *Heavy work*. Though bodily heat production while at rest is about 100 watts, it rises to over 1000 watts during severe exertion. Heavy lifting, moving or hauling involved in the handling of some materials therefore causes the occupant of a building to produce large quantities of heat energy which must be dissipated to the atmosphere, and a greater temperature difference between the body and the environment together with adequate ventilation enables this high rate of loss to be achieved.

So, the reason for lower temperatures in such industries is not so much that the designer can get away with them but that they are essential to the well-being of the occupants. At the same time, the heavy manual worker still requires a higher temperature in the mess room or other welfare areas where his energy output is lower.

A great deal of the heavy tasks in industry are now being carried out by mechanical means so the tendency will be to raise heating standards in order to maintain comfort, and therefore efficiency.

Special Applications

(i) *Operating theatres*. The highly specialized requirements of operating theatres are a study in themselves. Suffice it to say that the principal problem in maintaining comfort standards for the exacting tasks carried on is to keep the temperature down to reasonable levels while maintaining a good supply of clean air.

(ii) *Computers*. In recent years the limits of temperature and humidity for satisfactory computer operation have widened considerably owing to the widespread use of transistors in place of thermionic valves, and the replacement of magnetic core storage units by more efficient low energy cycling systems. As a result it is possible to site computers in normal rooms with an environment controlled only to the extent which is in any case desirable for human occupation. Many of the computer units have their own built-in forced ventilation system and, provided the room air temperature is within the normal range, require no further specialized treatment.

(iii) *Laboratories*. Laboratories can range from the heavy structural testing shed to the finely controlled high-energy physics research unit. As the requirements become more precise so the occupants are able to provide a more detailed brief. It is, however, an unfortunate truism that the more detailed the brief the more likely it is that it will be changed as time goes on. Provision of environmental control for these demanding areas should therefore take account of the probable need to alter the conditions and should be based on a wide range of selectivity for both air temperature and humidity, together with appropriately flexible ventilation systems.

(iv) *Process work.* 'Processes', in industry, generally mean chemical treatments to metals. They include pickling to remove surface deposits, etching to provide key for surface treatments, plating, coating, washing, rinsing and paint application. Some involve the use of acids, some alkalies, nearly all the tank processes are hot and consequently most involve fumes, which may be corrosive. The paint applications usually depend upon volatile solvents which may be asphyxiative or toxic. Obviously this is a demanding field where ventilation is of first importance, and the high rates of ventilation often necessary impose a heavy burden on heating systems. Many plant installations incorporate their own ventilation systems but the side effects of violent air movement on the near surroundings must be considered.

(N.B. One common factor in all these special applications is that of the continuous nature of their functioning. All may be required to work for twenty-four hours every day, and it may therefore be necessary for full stand-by plant to be available and reliable maintenance facilities provided.)

Importance of ventilation

It will have been noted that in each of the above occupational discussions the rate of ventilation has been mentioned as an important factor.

From previous chapters it will be clear that, when people are considered to be comfortable in relation to their environment, it results from a suitable combination of air temperature, radiant temperature, humidity and air movement. Such a combination can be called an 'Index of Thermal Comfort' and may, in its crudest form, be simply the air temperature. But, owing to the inadequacy of this simple index, more complex indices have been devised, such as the *Normal Corrected Effective Temperature Scale* (Chart 4, *Appendix* A). Such an index takes account of the problems we have already discussed concerning air temperature, radiant temperature and humidity, but up to this point air movement has been only sketchily covered.

Though certain of the statutory regulations state minimum ventilation rates, these tend to be somewhat incomplete. In consequence, ventilation is often overlooked in the design stage, but, fundamentally, adequate fresh air must be provided for the following reasons:

(a) To provide a continuous supply of oxygen for human functions.

(b) To remove products of respiration and occupation.

(c) To remove 'artificial' contamination arising from smoking, cooking and industrial processes, etc.

Considering the minimum ventilation rate when the density of occupation is known, the following schedule, Fig. 10.1, indicates acceptable levels. As it is often the case at the preliminary stages of design that the precise numbers are unknown, an estimate of probable occupation will have to be made. The function of a space can often give a good guide: for example, a press shop in a factory is occupied by large machines with large access gangways and few operatives, whereas a small component assembly-line may contain ten times the number of persons in the same area.

Air space/person		Minimum Fresh Air per person			Recommended Minimum if Smoking is Permitted		
m^3	ft^3	m^3/hr	ft^3/hr	Air Changes	m^3/hr	ft^3/hr	Air Changes
3·0	100	45	1500	15	54	1800	18
6·0	200	30	1000	5	36	1200	6
9·0	300	22	720	2·4	26	860	2·9
12·0	400 or more	19	600	1·5	22	720	2·4

FIG. 10.1.   Minimum Ventilation Rate where Density of occupation is known.

Since the height of the structure may not be determined at the early stages of design the cubic content may not be known. In this case it is possible to apply a ventilation rate to the floor area, ranging from $1·8 \, m^3/m^2$ hr ($6 \, ft^3/ft^2$ hr) to $15·0 \, m^3/m^2$ hr ($48 \, ft^3/ft^2$ hr).

Fig. 10.2 lists a number of statutory or officially recommended standards for air temperature and ventilation rates. If buildings of similar nature are met with, but no information is available, a judgement of suitable conditions can be made from the table.

Authority or Reference	Recommended Standards		
	Air Temperature		Ventilation Rate Air Change
	°C	°F	
Residential Accommodation			
for Elderly People —Local Authority B.N. (2) ..	21	70	3
Local Health —Local Authority Authority Clinics B.N. (3) ..	21	70	2
Junior Training —Local Authority Centres B.N. (4) ..	18	65	2
Adult Training —Local Authority Centres B.N. (5) ..	18	65	2
Factory Acts (1961)	20 (after 1 hr)	68	
Ministry of Education Bulletins Standards for School Premises Regulations (average)	18	65	2
L.C.C. Bye-Laws & Regulations			m³/hr ft³/hr
(Public Entertainment)	13	55	30 1000 per person
The Offices, Shops & Railway Act (1963) ..	16 (after 1 hr)	60·8	adequate
Hospital Building Notes (see *Notes*)			
I.H.V.E. (U.K.) (see *Guide*)			
A.S.H.V.E. (U.S.A.) (see *Guide*)			
M.O.H.L.G. Bulletins (see *Bulletins*)			

FIG. 10.2.   Some Recommended Temperatures and Ventilation Rates.

The rate of ventilation corresponds, or rather is related, to the speed of air movement across a space. The relationship is not direct, but depends upon the means of entry and exit, the size of the inlets and extracts and the complicated pattern of air currents. Flow will, however, be roughtly proportional to the rate of air change. Comfort depends upon a minimum rate being achieved, but also upon excessive speeds and draughts being avoided. Fig. 10.3 shows the relationship between the temperature of the moving air and its minimum speed for comfort.

Air Temperature		Speed of Air	
°C	°F	m/sec	ft/min
12	54	0·15	30
15–18	60–65	0·20	40
30	86	0·60	120

FIG. 10.3.   Acceptable Air Speeds at Various Temperatures.

Outside the United Kingdom different climatic conditions prevail. A brief indication of a number of different types of climatic ranges are illustrated in Fig. 10.4. It will be noted that, while conditions for Malaya and Iraq may be fairly represented by a group of figures, those for India or the U.S.A. are necessarily relevant to certain areas only, owing to the wide range of climates in those large countries. The U.K. figures are included for the purpose of comparison.

Country	Comfort Zone Effective Temp.		External Conditions					
			Winter		Summer			
					D.B.		W.B.	
	°C	°F	°C	°F	°C	°F	°C	°F
Malaya ..	23–25	74–79	7	45	33	92	28	83
Iraq ..	22–24	73–77	2 to 4	35–40	47	117	24	75
India ..	20–24	68–77	4	40	42	108	—	—
Australia	22–24	72–76	−1 to 7	30–45	35	97	24	75
U.S.A. ..	19–24	66–75 (s)	−24 to 2	10–35	35	96	24	76
	17–21	63–71 (w)						
Germany	18–23	65–73 (s)	−15	5	30	88	20	69
	16–20	62–69 (w)						
U.K. ..	18–21	65–71 (s)	−1	30	27	82	19	66
	13–20	57–68 (w)						

(s) = Summer   (w) = Winter

FIG. 10.4.   Some Recommendations for Design Temperatures
Overseas.

Air-Supply Requirements and Elementary Ventilation

Now that some of the elementary principles and applications of internal climatic control have been discussed and external climatic conditions considered, we can move on to the study of the means of providing the necessary heating or cooling required.

It could be said that cold air and its displacement is the basis of most heating applications. Air warmed by a heating element or surface rises to the ceiling of the space to be heated, displacing cooler air which sinks back to the floor at a point remote from the heating element, being then drawn across the floor towards the base of the heater to be heated up again. It will be seen that this circuit or thermal displacement of the air will cause convection currents, the air being cooled as it passes over room surfaces, giving up its heat to them and sinking as its density increases. The coolest surface in the normal space is the glass of the windows, and a rapid downward current of air always tends to occur near them. Thus, if a heating element is placed opposite a window, the two currents of air are complementary and a rapid convection is set up whereby cold air races across the floor, caused by the different density of the cold descending stream at the window and the vertically displaced warm air of the heater. This can result in varying degrees of discomfort and apparent 'draughts'. The correct placing of the heating element is therefore under windows or other cold sources as a first choice. This reduces the 'draught effect' within the room and produces a much more even temperature inside the room. If the sole heating element is an open solid fuel fire, the problem is difficult to solve. Firstly, combustion air must be provided for the fire, and this is best achieved by running an air duct from the external air to a point near the fire. (The best place, in fact, is in the cheek of the fireback, to provide both combustion air to sink on to the fire, and surplus air to keep chimney-draught going.) Secondly, to reduce the ingress of cold air at the window or doors, double-glazing well sealed becomes desirable, as does weather sealing at the doors. If a secondary heating element can be placed under the window to combat the descending cold air, a good standard of comfort can be attained.

When no duct exists to carry air to an open fire, the stack- or flue-draught caused by the displacement of smoke and hot gases up the chimney induces fresh air to enter the room through cracks and gaps in the construction.

Unless this demand for air can be satisfied by deliberate supply, the fire will not burn satisfactorily and will be highly susceptible to down-draughts in the chimney. Thus, draught-stripping a room can create the air-supply problem in respect of good combustion and minimum ventilation for occupants. The same arguments obviously apply to rooms housing boilers.

Warmed-air systems

In warmed-air systems, it is usual to design for a positive pressure throughout most of the building. This tends to reduce the effects of infiltration through the structure as well as preventing objectionable draughts, all draughts being outwards. To replace the air lost by outward draughts and by losses from kitchens, bathrooms, etc. (from which it is not normal to recover return air for obvious reasons), a proportion of fresh air is provided, ducted to the suction-side of the fan. The quantity of fresh air supplied varies, but 10 per cent of the output may be regarded as a minimum. This ensures that all the air is not recirculated, thus becoming stale and vitiated. Certain spaces are treated as expendable from an air-recovery point-of-view and include toilets, kitchens, process and plating shops (acid or caustic atmospheres) and certain laboratories.

External doors form a considerable problem in maintaining constant internal conditions. Each time a door opens and closes a sequence of pressure changes in the room causes interchange of air which can, particularly at a busy entrance such as that of shops, hotels, etc., unbalance the heating system completely. A common solution is to place a unit over the doorway (sometimes even under the floor) which discharges a curtain of warm air over the opening, so as to warm up the inrush of cold external air. Such warm-air curtains have the disadvantage that the air is necessarily at high temperature to be able rapidly to heat up the cold air, and also that the velocity required to obtain thorough mixing of the airs has to be high. Customers do not necessarily appreciate the difficulty of the problem. Strategic placing of radiant heaters can sometimes alleviate the effects without the need for an air curtain.

Mechanical ventilation

The provision of adequate fresh air in a building without relying on external openings is of course one of the basic problems of the modern building in temperate climates. Building Regulations require certain minimum standards of openable windows (unless other means are provided). Natural ventilation is therefore, to say the least, un-scientific and relies entirely on the human element for its success. Infiltration usually copes with the basic demand for oxygen but, as indicated previously with heat gains, etc., the internal atmosphere can become very stale and torporific unless a reasonable quantity of fresh air is injected.

The design of closely controlled artificial environments is therefore essentially one of ventilation. The addition of fresh air forms part of every ventilation or air-conditioning system associated with human beings and has a direct bearing on the heating or cooling plant, in addition to heat losses or gains from other sources. In fact, the infiltration of air entering a building is computed along with other factors when designing normal heating requirements. The unit of measurement for ventilation is normally expressed as 'air changes per hour', or can be stated as a volume per unit time (m^3/sec or ft^3/min), as in the previous heat loss calculations.

When the effects of natural ventilation are insufficient, these need to be replaced by powered ventilation. This can be achieved in a number of ways, the selection of method being determined by the nature of the problem and the function of the building and may range from window fans to large central-handling plant. The movement of air after discharge is complex and depends upon the entrainment of existing room air with incoming air. It is somewhat difficult to predict the paths likely to be followed by the air in a room, except in general terms. Thus, it is a matter of common logic to suggest that a localized input of air adjacent to a localized return-air outlet is liable to suffer from short-circuiting, and that a dispersed input of air remote from a series of air-return outlets is likely to result in good distribution. However, in the case of a laboratory, for instance, there is often a concentrated extract point at fume cupboards and, due to the overriding importance of maintaining a high rate of extraction from them, little or no other extract-air ventilation is provided. Such a case, incidentally, is one where it is important to maintain a consistent negative pressure in the room, too, so that all air tends to be drawn in, so as not to spread smells and fumes about the neighbouring rooms.

Apart from such special applications as theatres and laboratories, as above, there is a general tendency to use widely dispersed diffusers to spread small quantities of conditioned air at a large number of points, and to pick up return air at a lesser number of grilles which are still, however, dispersed about the space. In most cases both input and extract are at ceiling level, although, in the case of theatres, it is usual to supply at ceiling level and to extract at low level. In domestic installations, however, it is common to install the ground-floor supply diffusers at skirting level, or in the floor itself, served by ducts cast into the floor slab. The return air is released through high-level transfer grilles into a circulation space giving direct access to the return grille of the warm-air unit. The first floor can be the reverse of this, the supply air being ducted from the warm-air unit to the roof space and distributed there to ceiling outlets in the various rooms, the return air being allowed to return *via* grilles to the circulation space again. The disadvantages of this method are obvious. Privacy is impaired by the return-air openings, while floor outlets may be covered by furniture. A fully ducted return-air system, using the first floor to house the ducts, can overcome much of the privacy problems, and careful selection of outlet positions avoids the blanking-off problem (bearing in mind, of course, the main requirement of combatting the cold-air curtain at windows).

So far, mechanical ventilation has been dealt with as if it were only a matter of distributing treated air to spaces. This is an over-simplification of the problem but, basically, there are three combinations: low velocity, medium velocity and high velocity. These methods can be as complex as, for example, the dual-duct high velocity system, in which cold and hot air in separate ducts supply a special mixer unit (called an *attenuator box*) which can be adjusted thermostatically to admit varying proportions of cold and hot air to the room, at the same time reducing the speed of movement to provide acceptable noise levels.

Another type of system involves the discharge of air into a void over a perforated suspended-ceiling, at very low velocities, so that conditioned air is forced down into the room over the whole of its plan area. The velocities are extremely low but high rates of air change are obtainable, the return air – also being collected in the ceiling – giving rise to a negative pressure zone.

It is apparent, from the wide variety of methods available (which are discussed in detail later), that mechanical ventilation is not unduly complex but is deserving of careful consideration at the early stages of a design, since the movement of air involves much larger areas of ducting than pipes for water systems.

11: Elements and Zones within Buildings

HAVING CONSIDERED THE FACTORS influencing internal thermal environment, let us now consider the building shape and the zones and elements resulting from it. Specific occupational functions require a given space in which to be performed – e.g. to sit at a desk or drawing board, to operate a machine, to accommodate a machine manufacturing or producing an article, and space to assemble or store the articles, etc. Also determinable is the minimum height in relation to this plan area that will be acceptable functionally. For example, if it is only required to enable humans to move about within this area, the height could be scaled to this requirement. If, however, the objects within the space are larger or require moving from one location to another, the height will be governed by these demands. In this way, the optimum cubic requirements are arrived at. Beyond this point, the relationship of these volumes is governed by the interdependence existing between them, the availability of the site area and the economics of construction. It is at this stage that it is possible to analyse the thermal qualities of alternative building shapes and the element requirements.

It can be said that each building is an integration of large or small cells or units which, for convenience, can be called *Elements*. The thermal characteristics of each of these 'elements' can be expressed in terms of the ratio of *floor area* to *external cladding*. There may be a number of 'elements' within the building having different occupational requirements which, in turn, demand differing degrees of comfort. Hence, consideration of the element requirements and environment will establish the basic method of heating or cooling for 'elements' whose relative location determines the *zonal* requirements.

The resulting *zonal* requirements will establish the extent of the services distribution required and, in turn, the most rational location for the primary plant area.

The particular orientation of a building shape will produce zones due to the solar effects. For example, in the Northern Hemisphere north-facing buildings do not receive any direct sun, only re-radiation from ground or sky and, in consequence, the north face is always relatively 'cold' compared to the other faces of the building. In particular, the instantaneous effect of solar heating is more pronounced the larger the glass area, and it would be possible, by eliminating the glass area and designing a windowless structure, with an external cladding so calculated as to eliminate the direct solar effect, to produce a building with a high thermal inertia, thereby carrying over the effects of the solar gains into the late evening or night.

This serves to illustrate that *Thermal Zones*, whilst commonly the result of the diurnal changes, can be independent of them when they may be brought about by other requirements associated with the different 'elements' within a building.

This is more likely to be the case for buildings of the future, but, currently, the internal environment of certain types of building is virtually unaffected by external climate in the normally accepted sense and, in consequence, produces diverse problems of 'acceptable degrees of comfort'. For example, cathedrals, very large department stores and underground control rooms are but a few with very different thermal characteristics and diverse occupation requirements.

It is obvious, however, that the ratio of floor area to external cladding can only be used for buildings of similar occupational or functional requirements.

The two building shapes shown on the following page (page 116) and their relative orientation will illustrate the effects of thermal zones.

The 'T'-shaped building (Fig. 11.1) is assumed to be conventional in its layout, that is to say the maximum distance from external windows to corridor-line would be 6 metres (20 ft) with a corridor width of 2 metres (6 ft). The degree of the 'conditioning' could be a L.P.H.W. system of radiators or convectors, etc., with natural ventilation from openable windows. The probable extent of service ducts to meet requirements would be nominal.

If the building was to have a 'degree' of conditioning giving controlled temperature and humidity during summer and winter, a more detailed consideration of the elements and subsequent zones would be required. It should be obvious, also, that

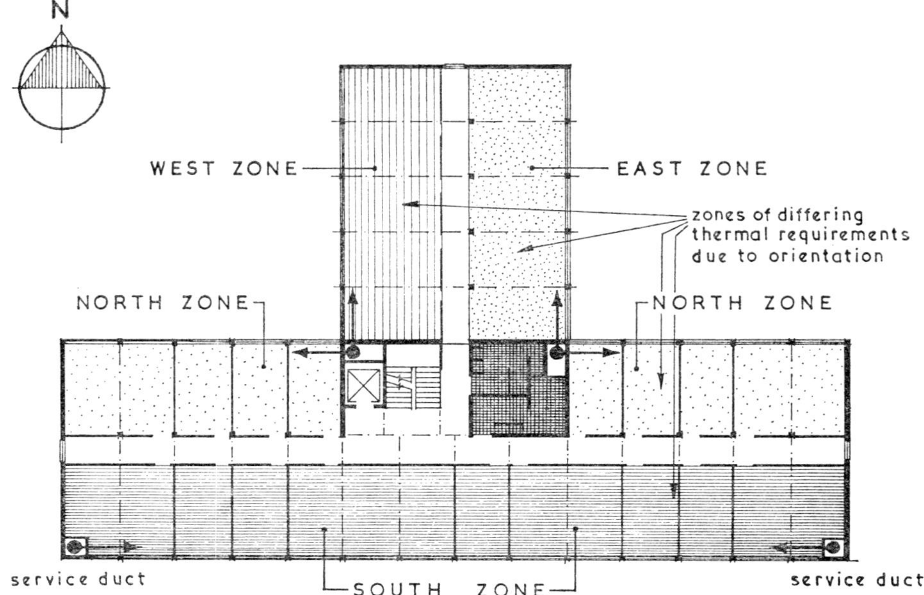

Fig. 11.1.   The Thermal Zones of a Typical Building.

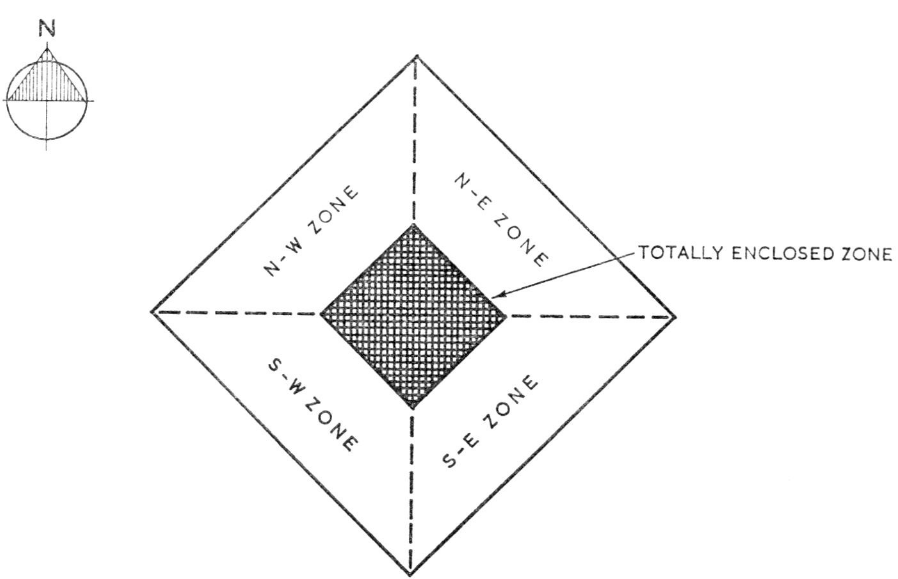

Fig. 11.2.   The Totally Enclosed Zone.

a higher proportion of duct area would have to be allocated for the same size of building.

The square-shaped building (Fig. 11.2), orientated as shown, would produce the pattern of zonal requirements indicated in the illustration.

If we assume the inner zone to be an occupied zone, the 'degree' of conditioning required would depend upon the number of persons, machinery and lighting or other miscellaneous heat gains and, on an intermediate floor, it would be virtually independent of external heat gains or losses. Each of the zones would again be comprised of a number of elements.

An 'element' may be defined as the smallest structure cell which, due to its location, orientation and functional requirement within a building complex, will subject a single human or a group of humans to different internal thermal environments within the building complex as a whole.

Let us consider, in Fig. 11.3 (*a*) and the accompanying details shown in Fig. 11.3 (*b*), an element which might be applicable to the building shape in Fig. 11.1.

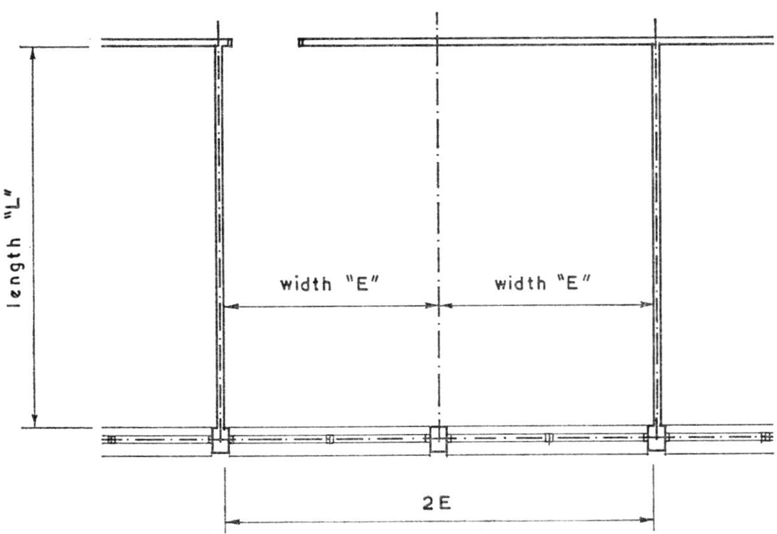

FIG. 11.3. (a)   A typical Element.

Height = H

2 No. Sub Elements 'E' = 1 Accommodation Element

Sub-Elements Dimensions	S.I. Units	Imperial Units
L	6·7 m	22 ft
E	1·75 m	5 ft 9 in
H	2·9 m	9 ft 6 in
Plan Areas	11·74 m	127 ft²
Glass Area	57% of wall cladding	57% of wall cladding
Heat Loss	1·4832 kJ/s or 1·5 kW (approx.)	5116 Btu/hr
t_o	−1°C	30°F
t_i	21·1°C	70°F
Heat Loss/cube	0·0436 kJ/s m³ or 43·6 watts/m³	4·26 Btu/cu. ft

FIG. 11.3 (b).   A Typical Element.

If, therefore, there are eight such accommodation elements with different thermal requirements due to orientation, it is possible to analyse each of these in relation to, say, natural lighting, and to adjust the ratio of $\dfrac{\text{Floor Area}}{\text{External Cladding}}$ knowing what influence this will have on the heat transfer and consequent heat loss for each of these. Then, by simply multiplying the numbers of such elements per floor, the gross thermal loss or gains for the building can be arrived at.

The advantage is that the gross heat requirement will be produced as the summation of all such element requirements, and

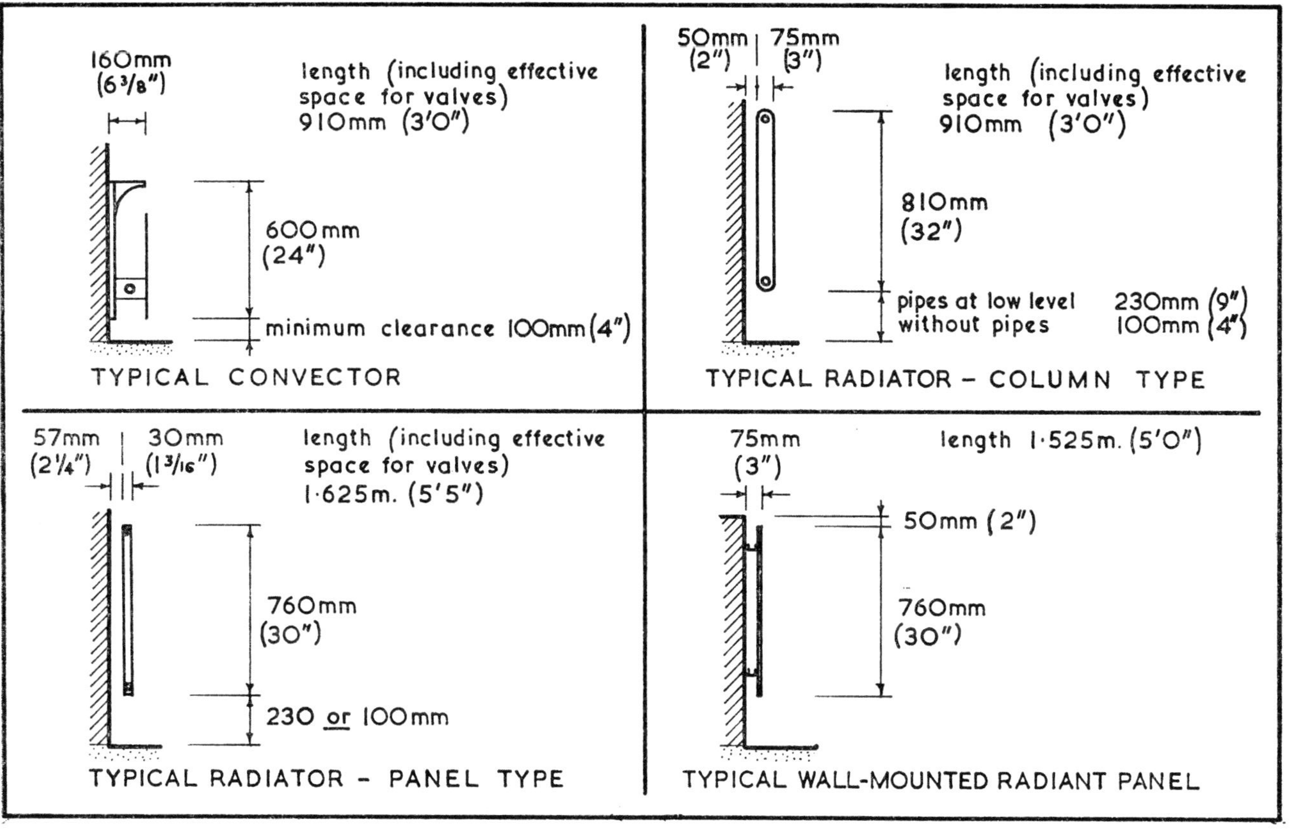

FIG. 11.4. Various Heating Units suitable for Element shown in FIGS. 11.3. (a) and (b).

it is simpler to assimilate and vary the technical details and requirements for an element than for the building as a whole. This approach is particularly helpful at the feasibility-study or preliminary cost-planning stage in which a number of variables can be applied to 'typical elements'. Summation of these will give an assessment of the overall problem without designing for the building as a whole.

For example, from the heat requirements for this location, it is possible to consider the alternative forms of heating unit that would meet these requirements and the respective physical dimensions. From this can quickly be seen the limitations imposed on the structure by the different types of unit, and hence, having reached agreement, it is established that this particular detail applies wherever the particular element may happen to repeat itself.

Typical applications of heating units to suit the element in Figs. 11.3 (a) and (b) are shown in Fig. 11.4 (page 119).

It is therefore possible to establish a fairly detailed appreciation of the building complex as a whole by consideration of these various elements. As indicated, the scale of the elements will be relative to the size of project – i.e. an element of a typical office block is the smallest cell that will enable a single human to carry out his duties within the space allocated; yet, on the other hand, the element can also be one building in a complex of buildings in which orientation and juxtaposition with other buildings on the site has to be considered.

Relationship of plant-space requirements

Consideration of the functional requirements of a building and the thermal characteristics resulting from this analysis, and influenced by the economic resources available, will establish the 'type' of installation that will be most appropriate. As the various plant areas are inter-related physically, they should be considered simultaneously in the initial planning of the building. Broadly speaking, there are three building groups which require services in the category of *basic*, *intermediate* and *complex*, as illustrated in Figs. 11.5, 11.6 and 11.7.

The arrangement shown in Fig. 11.5 can be considered as a traditional installation and applicable to smaller office blocks, schools and the like.

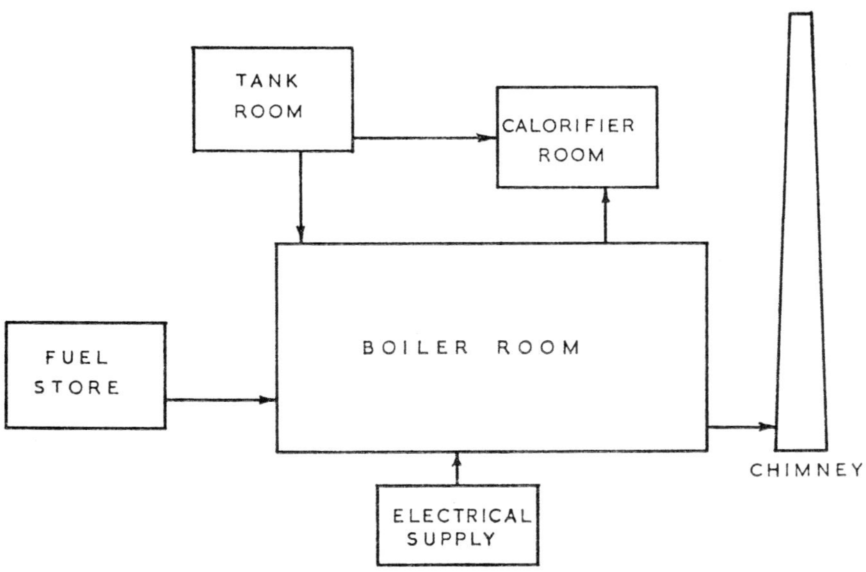

FIG. 11.5. *Basic* Plant Requirements. Conventional L.P.H.W. using opening windows for natural ventilation.

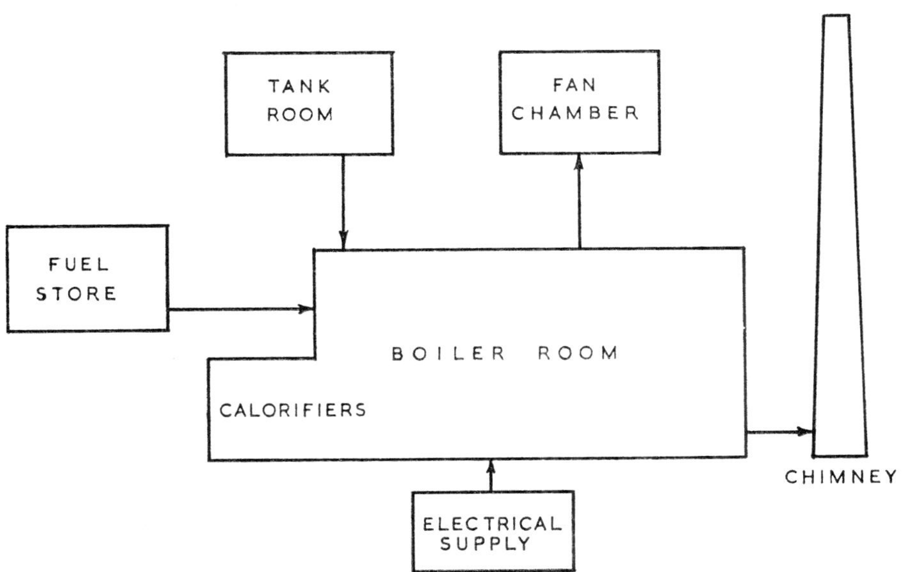

FIG. 11.6. *Intermediate* Plant Requirements. A combination of mechanical ventilation and heating.

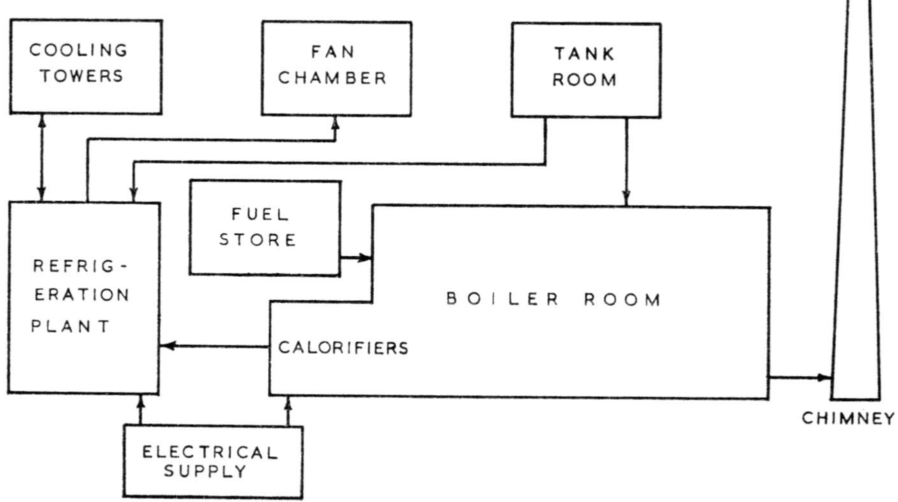

FIG. 11.7.   *Complex* Plant Requirements. Controlled conditions
of temperature, humidity and filtration.

The tank room, fan chamber and cooling towers are invariably
on the roof, but it is possible to have all plant housed on the roof,
dependent upon the choice of fuel and the local authority.

Fuel

The boiler plant, providing the heating medium, and the fuel
used have a considerable influence on the planning requirements
for the building and, as any of the four fuels is available, an
early decision must be made to enable the requirements to be
integrated into the building complex:

(i) Solid Fuel – coal, coke, etc.
(ii) Liquid Fuels – oil.
(iii) Gaseous Fuels – town gas, producer gas or natural gas.
(iv) Electricity.

The choice of fuel will depend upon:

(*a*) Limitations imposed on fuel handling due to site location.
(*b*) Economics of the cost of respective fuels.
(*c*) Types of fuels most appropriate to a particular building
and location.

Tabulated below are a number of common selections of fuels
according to these conditions:

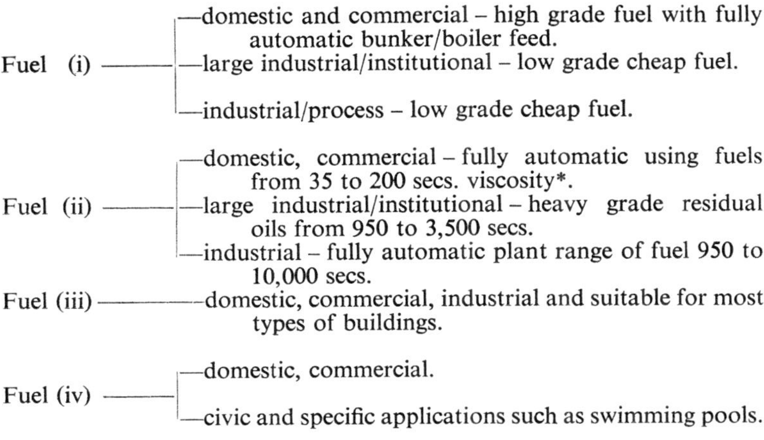

* (Viscosity of oil fuels is measured in 'Seconds Redwood')

Allocation of plant space

Fig. 11.8 serves to indicate very approximately the plant areas likely to be required for a calculated building cube:

Approx. Building Floor Area		Boiler Plant Floor Area		Fuel Store Floor Area		Boiler Plant Area as % of Building Floor Area	Fuel Area as % of Boiler Plant Area
m²	ft²	m²	ft²	m²	ft²	%	%
250	2,500	7	75	6	65	2·8	86
500	5,000	12	130	7	75	2·4	58
1,000	10,000	15	165	9	100	1·5	60
1,500	15,000	15	165	10	110	1·0	66
2,000	20,000	17	190	11	120	0·85	64
3,000	30,000	33	350	20	215	1·06	60
4,000	40,000	33	350	20	215	0·8	60
5,000	50,000	38	410	25	270	0·76	65

FIG. 11.8.   Approximate Plant Areas.

N.B. This table is based on a storey height of 3 m. (10 ft.) and is for guidance only, as area requirements vary according to type of plant and fuel used. Where a storey height is very different from the norm, the volume of the building should be related to an equivalent floor area of the normal storey height.

The permutations of type of plant and fuel are too numerous to draw definite conclusions on space requirements – but *always* allocate some space for plant. (Even if it is a wild guess it can invariably be 'moved' about, whereas if no space is allowed it always proves difficult to integrate it into the building plan.)

Boiler Rooms for buildings having a cubic capacity between 15,000 m³ and 150,000 m³ (500,000 and 5,000,000 cu. ft) can be considered at a sketch-plan stage as ranging from 0·1 to 0·08 per cent of the volume. Clear height within the boiler room will range from 3 m (10 ft) to 5 m (16 ft).

Fuel store: This is dependent upon the nature of the fuel and the storage capacity required, which is normally calculated for three weeks for medium-sized plant. As an estimate for early stage planning an allocation of 30 to 50 per cent of that allocated for the boiler plant may be taken, but it can be as high a proportion as 75 to 100 per cent.

Air-handling Plant: In an intermediate or complex services provision, additional plant space will be required for air-handling and cooling plant. (For details, see Chapter 14.)

Chimney sizes and heights: Once again dependent upon the type of fuel and the *Clean Air Act* of 1956, which covers boilers from 1750 kW (600,000 Btu/hr) to ten times this size, but there is a minimum of 13 m (40 ft) of 'uncorrected chimney height'.

At the early stages of planning it is often only possible to make a 'guesstimate' and, therefore, items to be borne in mind in considering boiler-plant rooms will be:

   (i) Access for installation and removal of plant. Minimum doorway 1·3 m × 2·1 m (4 ft × 7 ft).
  (ii) Adequate ventilation to outside air from 0·1 m² to 10 m² (1 sq. ft to 10 sq. ft).
 (iii) Noise – there is always a relatively high noise level from boiler plant, and its location where this does not constitute a nuisance-value is necessary. Alternatively, adequate acoustical treatment of the plant room should be considered.

The occupational requirements within a building will influence the size of plant areas; for instance, speculative developments or low-cost development might allow small margins on the plant installed, but it is essential in cases of hospitals, schools, hotels, etc., to provide stand-by plant to ensure that adequate services can be maintained at any time. Obviously, such factors influence the size and layout of the plant requirements and can only be assessed fully as the designs develop.

In conclusion, therefore, do not underestimate the probable plant areas required (even if this subsequently has to be modified), and plan them as integral parts of the building and in some relation to each area of particular plant, thereby establishing the necessity for service ducts linking the various areas.

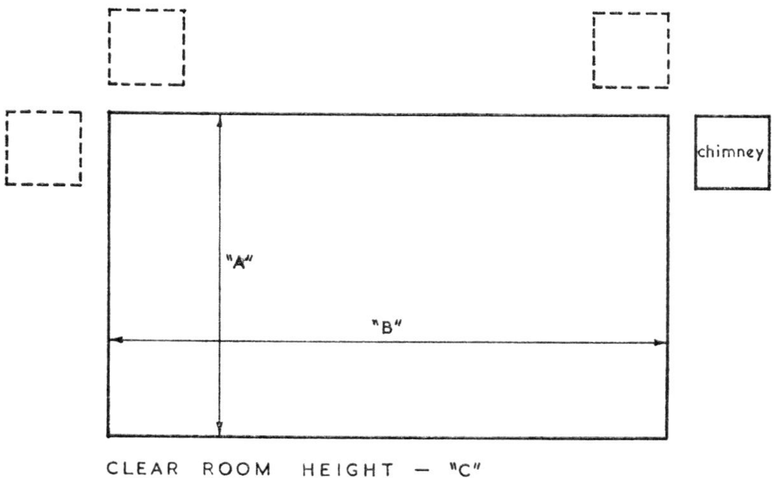

Volume of Building		Plant Room Dimensions						Approx. Chimney Area	
		m			ft			m²	in²
m³ × 100	ft³ × 1000	A	B	C	A	B	C		
0– 17	0– 60	2·4	2·7	2·1	8	9	7	0·05	81
17– 43	60–150	3·0	4·2	2·4	10	14	8	0·10–0·13	176–200
43– 59	150–200	3·7	4·8	2·4	12	16	8	0·17	260
59– 85	200–300	4·2	5·4	2·7	14	18	9	0·26–0·35	400–550
85–142	300–500	4·8	6·1	3·0	16	20	10	0·39	600

FIG. 11.9.   Plant Space for Small Installation.

12: Secondary Heat Transfer

General

NORMALLY, WHEN CONSIDERING THE HEATING of a building, one thinks mainly about heating the air within the building and, certainly, most systems are designed to do only this. It will be seen from some earlier notes, however, that the surface temperature of the fabric of the building influences the equivalent comfort effect felt by human beings within the space. Arising from this approach to the problem of comfort conditioning, the Romans used fabric heating for their public baths and for some villas in Britain. Heating the fabric, however, requires large surface areas at relatively low surface temperatures, to achieve satisfactory results. It is more appropriate to heat horizontal planes than vertical ones, due mainly to the disparity of suitable wall or vertically unobstructed surfaces available within a space. In consequence, the floor or ceiling are primarily used for fabric heating. The head being more sensitive when exposed to variable conditions of temperature and humidity, etc., radiant overhead heating can produce effects which induce subjective feelings of lassitude and headache if not properly designed initially. Transient spaces, however, such as entrance halls, lavatories or corridors can be so heated with advantage, as in fact can laboratories, hospitals and offices. When combined with mechanical ventilation, the result can be extremely effective in terms of comfort.

Fabric heating: underfloor

Two principal systems exist for underfloor heating: electric-cable networks and embedded-pipe coils. The electrical systems use a sheathed resistance wire, the cables being either cast direct into the screed or drawn through conduits to provide re-wireability. The network for each space is separately calculated on the basis of heat loss, and it is good practice to provide a higher density of wiring at external walls so as to combat the cold air currents at windows. In a fixed-plan building each room is separately wired, supplied and fused and may incorporate a thermostatic control. In flexible-plan buildings, such as office blocks or factories, the wiring is arranged in standard sections and may be so planned as to give an overlap of different circuits (subject to the planning of the supply) so as to provide variability of heating rate. Very few electrical underfloor systems use normal tariff electricity, however, as the running cost would be too high. The systems work on a night-rate (off-peak) supply whose timing is controlled by a time-switch set, and sealed by the supply authority. Off-peak electricity tariffs often include a day-time off-peak boost to top-up the system. From this it will be seen that electrical underfloor heating depends upon the heating-up of a large mass of concrete, so that heat is released by that mass slowly throughout the day. The system relies on the thermal inertia of the building discussed earlier, and is subject in certain circumstances to a slow response and lack of control of output. If the building requires to be intermittently heated, this system would not be applicable. Continuous occupational requirements would, however, be considered a suitable reason for such an installation. The electrical system is fairly cheap to install, moderate in running cost and low in maintenance. No plant or flues are required and no fuel storage is involved. These factors, coupled with the absence of supervision permitted by the system, make it a popular choice for flats (often only as back-ground heating) and for houses (backed up by other local heating units) and certain types of offices.

There are other proprietory electrical floor-heating systems utilizing thick timber floorboards with cables run in grooves in the edge of the boards. This makes it suitable for installation in the older houses, though the dangers of making floor fixings with such shallow protection to the cables must be borne in mind.

The other method of floor warming uses 'coils' of small-bore pipes buried directly in the floor screed, or buried within an asbestos sleeve within the screed. Low-pressure hot water is circulated through the pipes, the prime unit being a hot-water boiler using any of the normal fuels available. The system can be controlled as required for the particular external climate conditions using a mixing valve and circulating pump.

Floor temperatures should be within the range of 20°C to 27°C (70° to 80°F), the primary-flow temperature being at approximately 43°C (110°F) for directly embedded systems, and up to 80°C (180°F) for sheathed-embedded systems of under-floor heating.

Constructional considerations

Having looked at the methods available for floor warming, it is necessary to consider the implications of this as far as building construction is concerned. Firstly, since the aim is to warm the inside of the building and not the external ground, it is clear that conduction paths from the slab to the exterior must be broken by insulation; secondly, the installation of cables or pipes in ground-floor slabs must be protected from damp, especially as the moisture content of the slabs will be very low and therefore highly absorbent; and, finally, the provision of floor finishing with high thermal resistances should be avoided over the heated floor.

Some particular points to bear in mind are:

(*a*) Due to the temperature of the slab, timber-floor finishes are liable to shrink unless laid at a low moisture content. There is, however, always a delay between laying the floor finish and turning on the heat. During this time, wood blocks can absorb moisture from the atmosphere and may buckle. It is therefore important to choose a medium moisture content, say around 8 to 9 per cent, so that, on further shrinkage, small cracks open-up between the blocks. When heating is turned off, the blocks slowly absorb moisture and have a little room to expand.

(*b*) Heat absorbed by the floor slab will tend to build up under floor-mounted furniture, thus where permanent fixtures are to be mounted directly on the floor it is advisable to leave out a section of the heating grid. When it is considered that the usual warmed floor temperature is in the 25°C (75°F) range it will be seen that food storage cabinets and antique furniture could suffer from lack of attention to this factor.

Carpeting, particularly when laid on cellular underlay, forms a good insulant, preventing normal heat dissipation from the floor. This can raise the surface temperature of the screed or block beneath the carpet to much higher levels than the design range, and can aggravate the desiccation and shrinkage problems referred to above. Alternatively, the floor temperature must

PLATE I. A range of Column Radiators in an office building (*see page* 131).

PLATE II. Unit Heaters (side-throw) in a factory (*see page* 131).

be designed-down to suit the specific critical temperature of overlay materials, and this in turn probably means that supplementary heating will be required in addition to the floor-warming system.

Floor warming in multiple occupancy building raises the question of heat loss to adjoining occupancies – in this case, the storey below. It is therefore necessary to provide an insulating layer beneath the heated screed. This can very satisfactorily be the impact-noise resistance layer, too, and bonded-fibreglass quilts or precompressed expanded-polystyrene boards will perform both duties.

In order to avoid inequalities in the consumption of heat by independent users, it might be appropriate to allow for a fixed minimum input of 10 per cent to attempt to reduce the heat loss between independent users.

Fabric heating: ceiling

Ceiling heating, which normally uses low-temperature hot-water embedded-pipe systems, falls broadly into two categories, namely:

(*a*) Embedded pipes directly within the main structure.

(*b*) Independently installed as part of the secondary interior finishes of a building.

The embedded radiant-heating systems of category (*a*) have similar thermal characteristics to the embedded floor panels; the response can be somewhat quicker and higher operating surface temperatures are permissible, depending upon room height.

The independent system of ceiling heating is usually an integral part of a proprietary suspended ceiling where the heating pipes form part of the ceiling-panel suspension; or, alternatively, if electrical, the cables can be laid directly on to the ceiling panels. In both instances, an insulating blanket is laid on the back of the heating grid. Such systems offer the advantages of an acoustic-heated suspended ceiling. The thermal characteristics of this system, without the direct influence of the structural mass, give rise to quicker response to thermostatic control. Dependent upon mounting heights, such systems can be operated at higher temperatures without danger of structural damage.

In certain circumstances, these ceilings can be used as heated/

cooled/acoustic ceilings for comfort conditioning, as discussed in a later chapter.

Unit heating

By far the most usual form of heating is the use of local heating units supplied from a central source of heat energy, such as a boiler. 'Central Heating' as a term implies this, and, strictly speaking, would be considered to exclude systems relying on the distribution of fuel energy, either gas or electricity. However, it is more appropriate to regard central heating in the broadest sense as including all forms of secondary-heat exchange supplied from a central source. The manner of dissipating heat energy to building spaces is determined by a number of factors – some practical, some aesthetic, some economic. The practical factors include the complexity of the spaces to be heated which can demand full control over the heating in each space or zone; and the form of the primary heating medium or fuel, etc. The aesthetic factors usually relate to the desirability for the concealment of distribution and the capability of blending the heating elements as unobtrusively as possible into the decorative scheme for the space heated. The economic factors are three-fold: first, the installation cost; second, the running cost; third, the maintenance, replacement and depreciation costs.

Pipe Systems

The earliest central heating was, of course, the Roman Hypocaust, using warm air. Subsequently, systems relied on the gravitational circulation of heated water around large-bore pipe circuits, whereby the pipes were capable of dissipating the required amount of heat. The pipes had to rise and fall correctly to create the requisite thermo-gravitational slopes to assist circulation. They were somewhat inflexible and obtrusive and are rarely used today. Steam was used in a similar simple way, the pipes being much smaller since the energy potential of steam is many times greater than that of an equivalent weight of water. These were limited to the larger industrial applications. The pipes, in fact, were the heating units. Finned elements were added to increase the heating surface and rate of heat dissipation.

Radiators

The early cast-iron radiator developed as a water or steam container having a relatively large surface area capable of dissipating heat at a more localized point. The term 'radiator' is, however, rather misleading in that the proportion of radiant heating emitted is only about 30 per cent, the remainder being by convection.

The highly convoluted surface of the column radiator is much more efficient (in the terms of space occupied by the unit) than the more simple panel radiator; but here an aesthetic choice is involved, and a change in ratio of convective to radiant-heat output also results (see Plates I and II, facing pages 128 and 129). There are numerous forms of other radiant/convective heaters, such as the oil-filled electrically heated radiator, the continental finned type and the electric storage-heater type.

Convectors

Having appreciated the high proportion of convected heat produced by radiators, it is a small step to encase the radiator element or finned tube, leaving an air-entry space at the bottom and an air-exit space at the top, to produce a unit being mainly convective. Aesthetically, the convector has much to commend it, and it can be free-standing as a single unit, or continuous in the case of sill- or skirting-type units.

Continuous Units

When individual radiators or convectors are used as the heating element, each unit can be separately controlled by valves giving a high degree of local control, but, when continuous convectors and skirting heating are used, the heating element is, like a pipe ring, a continuous part of the circuit. It cannot therefore be valve-controlled, except as a complete circuit or branch of a circuit. A certain degree of control of the output from these is achieved by controllable damper-blades which shut off the exit grilles, preventing convective movement of the air, reducing output down to as little as 10 per cent. Clearly, the heating element will dissipate a small quantity of heat to the casing, which will radiate it to the room; but the quantity is small and may be disregarded for all practical purposes. Since the casing becomes heated by the element, it is clear that the wall on which the element pipes are mounted will also gain heat, so it is advisable to provide an insulating layer of fibreboard, or similar material, between the casing and the wall. It is inadvisable to use direct gas-fired radiators or convectors unless there is adequate fixed ventilation to reduce the effect of waste gases; preferably, balanced-flue type units should be used. At a domestic level, the significance of seeing a unit 'glow' is still important, and hence many convectors also incorporate red or amber lamps in the casing to achieve this, while some also have a radiant bar to supplement the convected heat, so producing the combined radiant/convective unit.

Radiant Heating

The need for a proportion of radiant heat to achieve a balanced level of comfort has already been referred to, but some systems rely almost entirely on radiation, much as the low-temperature floor or ceiling methods do. The use of high-temperature radiant methods such as gas fires, electric bars and overhead gas radiant heaters has to be carefully considered, due to the effects of the localized source in that the immediate zone of the heater can become uncomfortably hot, to the point of scorching, while the remote zones are too cold. The heating, in consequence, is liable to be 'patchy'. Also, one unfortunate feature of direct electric radiants is that the direct impact of the full range of wavelengths on the eyes can cause irritation.

Radiant heating by piped hot media is somewhat different in that no luminous waves are produced, the surfaces being capable of emitting only the lower-frequency radiation at the temperature used. Generally speaking, a low-temperature hot-water system of radiant-panel-type heating is somewhat restricted, as the areas of radiant surface required become very large and ineffective above 3 m (10 ft), but for industrial and commercial heating, where medium- or high-pressure hot water or steam are available, effective heating is achieved by means of high-level radiant heating. 'Low temperature' is a relative term implying, in this case, a range from 93°C (200°F) to 150°C (300°F) as against the 537°C (1000°F) of a radiant electric bar fire. Clearly then, even these 'low' temperatures are such that the elements must be mounted so that people cannot come into direct contact with them. The usual method in factories is to install a number of parallel high-temperature pipes below the roof structure, with metal-backing panels of various shapes which, in turn, are insulated with foil and quilt above, so that the face of the panel becomes hot and, in turn, radiates this heat downwards. A similar principle applies to a radiant-panel system which uses high-temperature air in a closed circuit of fan and heater, instead of high-temperature hot water or steam.

The wall-mounted low-temperature radiants follow much the same pattern as the overhead units but must have a lower surface temperature if they are within reach of occupants. The radiating panel, therefore, is mounted in front of the heating pipes and the back of the unit is formed of reflective and quilt insulation. The contact between the hot pipes and the radiating panel is dispersed over a number of points so that no 'hot spots' develop. The surface temperature of such a panel should not exceed about 60°C (140°F), so it follows that, with a lower temperature differential than the overhead radiants, a larger surface area is required to achieve an equivalent heat output.

Low-temperature electric radiant panels follow the same pattern as the piped water or steam panels and are available in small units for high-level mounting in small rooms. They are not susceptible to damage, are easily installed and are low in maintenance cost, this being ideal for isolated spaces.

The advantages of radiant heating for loading docks and entrances have been referred to, but it should be borne in mind that radiant heat, being a wave-form, can be interrupted by obstructions and is not therefore suitable for storage areas where racks or stacks of goods may blank-off areas, unless well-defined gang-ways are known to exist. Other types of element heating units are in use, such as the electric tubular heater, which is usually an elliptical tube with electric resistance elements bedded in insulating material within the casing. They dissipate up to 80 watts per foot run and are very simple to install, though they are expensive to run off normal daytime supplies due to the tariff charges. They are therefore usually confined to spaces requiring spasmodic heating at low capital investment, or are used as occasional boosters to assist other forms of heating system.

Storage Heaters

To take advantage of the cost of using electricity at a reduced tariff, the off-peak thermal storage-heater was developed. In its simplest form it consists of a high-density material which has high capacity for heat absorption and releases this heat slowly *via* an insulated jacket, after the supply is cut off. The thermal inertia of such heavy materials is the basis of the block storage used for this form of heating element. The method of control is achieved by incorporating, within the insulating jacket, adjustable flaps to permit the unit to operate as a convector. A more sophisticated form of block storage heater uses a fan-assisted ducted system in which the fan is thermostatically controlled to produce an 'on-off' condition of heating.

The circuitry which serves off-peak heaters (and electric floor-warming systems) is quite separate and distinct from the general building, lighting and power-supply system. It is controlled by a time switch provided by the Electricity Board, the time unit being sealed to prevent tampering. A thermostatic control regulates the quantity of energy put into the storage units and automatically cuts out when the system is full.

Fan-mounted Convectors

Conversely, some methods of heating depend entirely on convection, ranging from the natural-draught convector, already mentioned, to the fan-assisted convector. These units can be mounted at high or low level, free-standing or recessed, and, when appearance can be at an 'industrial' level, these units are referred to as 'unit heaters'. There are two criteria, apart from appearance and cost, which have to be considered. These are: (*a*), the suitability of the noise level which is associated with the type of unit; (*b*), the effective height available or required for mounting these units – the greater the mounting height, the greater the throw required from the unit, with increase in fan power and noise. (See Plate III, facing page 144).

The fan-assisted convector can become, in effect, a medium to large air-handling plant usually associated with ducted distribution off the unit. At or beyond this level, the 'sense' of element heating changes, and consideration should be given to the use of such units as a central-station plant in which the primary-heating medium is ducted warm air.

Warmed-air heating

In the heating methods discussed so far, each system takes the heat energy to the space in some concentrated form, whether hot water, steam, gas or electricity, and raises the air temperature within the space. In considering warm-air heating, a quantity of air in excess of that in the space is introduced at a suitable temperature; the excess air is then either re-circulated from the space back to the unit or allowed to dissipate through natural or artificial outlets in the building construction. Whilst the temperature of the air is automatically raised by entrainment with the incoming ducted air, it also gives a distinct air change within the space due to the displacement by the incoming air.

It is obvious that, by design, different ratios of $\dfrac{room\ air}{incoming\ air}$ can be achieved and, when mixed with a proportion of $\dfrac{recirculated\ air}{outside\ air}$, control of air movement is achieved to give specific air changes within a space and controlled rates of ventilation using outside air.

There are two elementary systems of supplying air in a ducted system. They are: (*a*), the traditional low-velocity system, with main duct velocities up to 5 to 7·5 m/sec (1000–1500 ft/min), requiring no attenuation on discharge into the heated space; (*b*), the high-velocity system, operating at up to 20 m/sec (4000 ft/min), for which system noise-attenuator boxes are required on discharge into the room or space. The design of efficient air systems must take account of the principal problems of the medium itself.

Firstly, the air as a heating medium has a relatively low capacity to absorb and store heat, unlike water, and in consequence must be moved in large quantities if equivalent heat output is to be achieved. Secondly, air in circulation is not pure: it contains minute dust particles to a greater or lesser degree. Thirdly, as explained earlier, air is subject to complex movement patterns, due to its specific density at different temperatures, which results in its taking the line of least resistance once it is unrestrained within a space.

In ducted-air systems, the fan is the equivalent to the hot-water pump in distributing heat about buildings and, due to the dynamics of air movement and the power associated with it, noise and vibration have to be taken into account.

The ducted warm-air system can be simply plenum ventilation in which the excess air within a space is discharged to atmosphere, or a balanced system in which the recirculation air, supply air and fresh-air intake are balanced against each other. In moving any fluid, liquid or gas, contained within a pipe or duct, there is a resistance to flow, and it is the sum total of resistance of the supply and return ducts, together with the volume of air to be moved per unit time, which determines the fan type and size. Three types of fan are used:

(a) *Propellor Fan* – a simple type of blade-fan suitable for only low-resistance systems, i.e. for small lengths of ducts without ancillary equipment.

(b) *Axial-Flow Fan* – an aerodynamically designed fan whose blades and mounting are such that it can operate against a high resistance, is compact, but normally requires noise-attenuation filters.

(c) *Centrifugal Fan* – a drum-shaped impeller rotating in a scroll-shaped case, whereby air is spun at speed so that centrifugal force throws it outwards. It breaks free of the drum where the scroll widens out and is driven along the duct. This type of fan can operate against high resistances and is suitable for most air-handling problems associated with comfort conditioning. It does, however, require a reasonable amount of plant space.

The supply air is heated to the designed level *via* a heat-exchanger called a *heater battery*. The heater battery can use either steam, hot water, electricity or a direct heat-exchanger using a gas- or oil-fired furnace.

The heat transfer is achieved *via* the boiler, producing heat energy in the form of hot water or steam which is circulated through the heat-exchanger. Air flowing through the heater picks up heat and is discharged through the fan into the duct system. The flue gas/air exchanger is used when a furnace is employed as the direct heating medium. The furnace gases pass over the heat-exchanger and the ducted air again picks up heat in a similar manner to the ordinary heater battery.

Air filtration

The 'degree' of air cleanliness required depends upon the level of activity within the space and can vary from coarse filtering to high-efficiency filtering. There are approximately six methods of filtration through this range. In certain installations, such as those for industrially 'clean' processes, a number of stages, each progressively more efficient than the last, are used, but for normal installations a simple filter of disposable pattern will suffice. Simple filters use fibreglass, or fabric and fibrous materials mounted in light perforated frames, and withdrawable for cleaning or replacement. Different grades of filter will deal with varying particle sizes, but it is natural that the finer the filter the higher the resistance to air flow. There is, of course, a low limit to the size of particle which can be trapped by a given type of filter. For average use, these are capable of enmeshing down to 10μ diameter particle size. This is dependent upon the speed of the moving air and type of filter medium. It will be evident that any interruption to the cross-section of the air flow constitutes a reduction in duct size; hence, to pass the same quantity of air in unit time, a filter requires to be made a good deal larger than the equivalent size of the duct (carrying the same quantity of air) which has to be expanded locally to accommodate it.

The adhesion of dust particles to dry fibres varies according to the nature of the dust, and it is therefore a fairly obvious step to coat the fibres with some sticky material to ensure good retention of the dust impinging on them. The most common form of viscous filter is the oil-coated mesh. The mesh is often metallic wool formed into a thick mat and dipped in light oil. The filter pads are 'washable' and can be re-used over a long period. The viscous filters vary widely in their capacity to handle dust, but can retain particles down to about 5μ diameter.

Dust particles can be given a static electrical charge by electrical equipment in the system. If, therefore, a plate of opposite polarity is introduced into the air stream, the dust particles will be attracted to it. This is the general basis of another form of filtration known as *electrostatic precipitation* which is capable of filtration down to $0{\cdot}01\mu$ diameter size. The size of particle is not in itself sufficient to categorize filter performance and must be related to a gravimetric efficiency, or equal, as called for in B.S. 2831 or B.S. 1701.

Associated in some instances with dust removal is odour removal, and absorption units of activated carbon or similar treated materials will achieve this. The cost is comparatively high, but may have to be accepted for certain specialized problems.

Humidity Control

When using a ducted warm-air system for heating, it is normally necessary to increase the humidity of the air in winter to avoid excessive dryness. This can be done by various types of air washers, such as the mechanical injection of moisture vapour into the air stream, or evaporation of vapour into the air stream.

In humidifying or de-humidifying, a spray-type or capillary-type air washer is sometimes used and is in itself effective in acting as a primary filter medium, but it is always used in conjunction with one of the other forms of filters with higher efficiencies.

Secondary heat transfer (H and V systems)

Having covered the principles and some general architectural aspects of heating systems, it is necessary to consider the various types of appliances, heating media and distribution required for an installation in further detail, to complete the picture.

Basically the types of heating appliances are as shown tabulated below:

Description	Types	Application	Material
(1) Radiators	(a) column	— commercial, industrial	Cast-Iron
	(b) panel	— commercial, domestic	Mild Steel Copper
(2) Radiant Panels	(a) high temperature	— industrial	Mild Steel
	(b) medium ,,	— commercial	,, ,,
	(c) low temp. floor or ceiling	— commercial, domestic	Copper, Mild Steel, Polyurethane
(3) Convectors	(a) natural	— commercial, domestic	Mild Steel
	(b) fan assisted	— commercial, domestic	,, ,,
	(c) unit heaters	— industrial	
(4) Ducted	(a) low velocity	— domestic, commercial	Aluminium Mild Steel
	(b) high ,,	— commercial/ industrial	PVC Glass Fibre Lamination

Heating media. There are four basic forms to be considered and the diagram below will indicate the limitations of direct application of the systems, scheduled above, with particular primary-heating media:

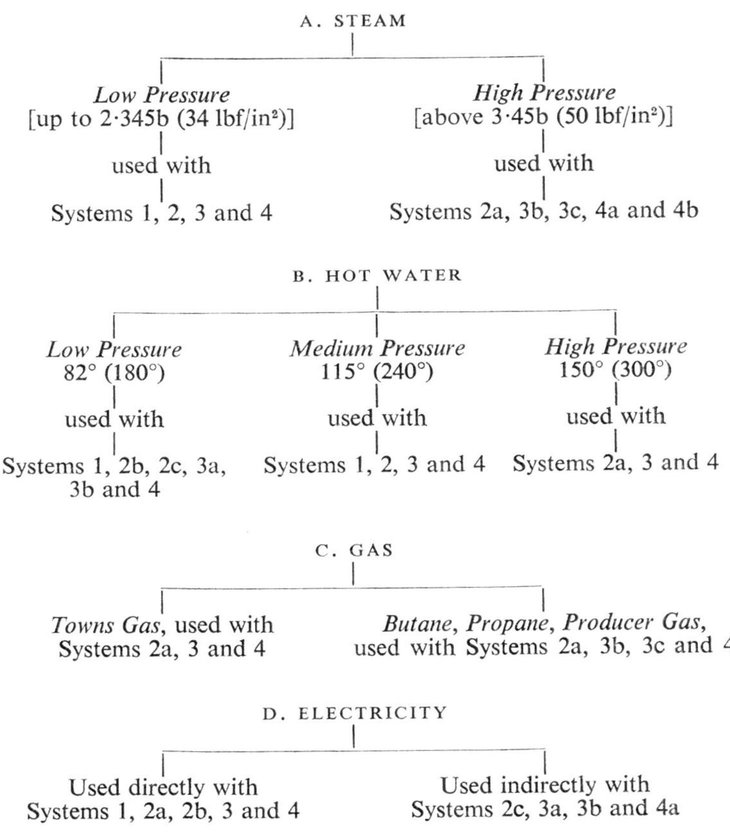

It is possible and practical with the high-pressure hot-water and high-pressure steam installations to 'break-down' the primary-heating media to suit all, or most, of the types of heating. This is done by using heat exchangers, such as steam/water or water/water calorifiers, thereby achieving the maximum heat potential to be distributed in the higher temperature and pressure medium, before breaking down to the lower degree of temperature and pressures acceptable for most systems. It is not practical or economical to eliminate the heat exchanger by using a mixing valve, as is the practice with M.P.H.W. primary heating media, because of the necessity to subject all the secondary heat-transfer system to the pressure of the primary supply, even if the system could be operated at the lower temperature by mixing. Both of these practices are adopted when dealing with district heating installations. The choice of primary operating temperature and pressures is dependent upon the size of the installation and/or the distances over which it is designed to transmit heat.

The use of direct gas and electricity with systems or appliances is covered in the diagram, but, obviously, being also primary fuels, they can be used indirectly to fire primary heat generators to produce either of heating-media A or B.

The methods of *distribution* of heating media (excluding gas and electricity) are as follows:

Steam

Steam depends upon a change of state from water to vapour and back again to water. The vapour, being under pressure, requires no mechanical assistance and relies on thermal change to provide movement.

Water

Water is still a cheap and easily obtainable fluid, its limitation being that, in order to use it without changing its state from a liquid to a vapour, it is necessary to keep its temperature below 100°C (212°F) at atmospheric pressure. This is done by using a feed and expansion tank and is said to be an 'open system'.

To increase its heat-carrying capacity it can be raised in temperature by having a 'closed system' which, in turn, will develop steam or high-temperature hot water according to the pressure at which the system is working. The circulation of the water, in very simple cases, relies on a thermo-dynamic movement due to the different density of the fluid at different temperatures. Generally, the water, as a heating medium, is circulated by means of pumps which range from the in-line immersed motor circulator of the domestic installation to the water-cooled direct or indirect-driven distribution pumps of the large installation working at high temperatures.

Hot Water: Methods of Circulation

Gravity Circulation (1) (i.e. thermal displacement due to different density at different temperature)	*Pumped Circulation* (2)
Used for L.P.H.W. only	Used for Low, Medium and High-Pressure systems

Disadvantages:
  Large pipes.
  Limited arrangement of pipe-work to ensure good circulation.
	Advantages:
	  Small pipes due to rapid circulation.
	  Flexible arrangement of pipe-work.

Ducted Air

This is included as a method of distributing heat media but, by comparison with the other media, is limited in its application to simple buildings.

There is no charge (as yet) for the use of air as a heating medium, but it is more often used in conjunction with a primary-heating medium through a heater battery of either steam or hot water. Currently, however, the use of air with a direct oil, gas or electric heat-exchanger is becoming more frequent, particularly in domestic and industrial applications.

The limitation is on the temperature at which the air can be controlled, i.e. at temperatures above 50°C to 60°C (120°F to 140°F) it is difficult, if not impossible, to keep the air down to an acceptable level within a space into which the air is supplied.

The use of 'high-temperature' air in a closed system with direct gas- or oil-fired exchanger is used, but the heating effect is achieved by converting into a radiant panel system and is used for industrial heating applications.

The two basic methods of distribution (excluding the closed cycle air-heating system) are either:

(*a*) traditional low velocity distribution with maximum velocities up to 7·5 m/sec (1500 ft/min);

(*b*) high velocity systems with velocities up to 20 m/sec (4000 ft/min). In this application it is necessary to provide acoustically treated outlet boxes.

Pipework distribution

Steam is primarily for industrial use and there are always two pipes: a Flow (steam) and a Condensate Return (an exception is a vacuum system used primarily in the U.S.A. requiring *one pipe* which provides for both steam flow and condensate return).

Hot Water. There are a number of basic methods common to most applications broadly described under the headings of 'one-', 'two-', 'three-' or 'four-pipe' systems. The first two are more commonly used for internal networks and reference to any standard works will show these systems diagrammatically. The last three systems are more applicable to the external distribution associated with thermal network systems.

<table>
<tr><td>L.P.H.W.</td><td>M. & H.P.H.W.</td></tr>
<tr><td>(a) One-pipe systems</td><td>(a) Two-pipe systems</td></tr>
<tr><td>(b) Two-pipe systems</td><td>(b) Four-pipe systems</td></tr>
<tr><td>(c) Combination of (a) and (b)</td><td>(c) Three-pipe systems</td></tr>
</table>

13: Primary Heat Generation

General

WE HAVE DISCUSSED THE VARIOUS FORMS of dissipating heat to building spaces but have not yet considered the various forms of primary heat generation.

Let us consider an elementary prime-heat generator – namely, the back-boiler. This is used only at a domestic level and consists of a simple cast-iron or steel vessel in which water is contained and heated by being located at the back of a fire-place subjected to both direct radiation from the fuel and convection of flue gases passing over it.

The principles of this are extended to give a variety of types of boiler for different required outputs. The main variation is whether the hot gases pass through tubes or flue ways which, in turn, are surrounded by and heat the water, or, alternatively, the water is contained within the tubes or water-ways and then heated by radiation and flue gases from the furnace. The two forms can be designated simply as:

> (*a*) flue-way type boilers;
> (*b*) water-way type boilers.

The materials used for the various types of boiler can range from cast-iron or steel-sectional boilers or steel shell-and-tube flue-way type boilers, used for small to medium-sized heating installations, to systems using steel-tube water-ways, erected on site and encased in fire-brick with outer metal-casing, for medium and large installations.

In all of these boilers, water can be heated within the limits of the materials used from low temperature and atmospheric pressure [82°C (180°F) and 1 bar (14·5 lb f/in²)] to high temperature [180°C (360°F) and 12 bar (180 lb f/in²)].

Whenever water is heated, it expands, and the increase in volume due to expansion must be accommodated. This is done within the feed and expansion tank used with the 'open' system that is working up to 93°C (200°F) and atmospheric pressure. A special type of ball valve is used in preventing overfilling of the tank when the system cools down and the water contracts again.

In a steam boiler, the water contained is heated beyond boiling point, the whole system being sealed so as to permit the pressure of vapourization inside the steam space to build up to a pre-determined limit. The steam flows out through the flow pipe, circulates around the heating system, cooling as it goes, and finally condenses back to water. The condensate is collected and pumped back to the boiler. The system depends upon the arrangement of steam traps maintaining the steam pressure within the system, at the same time permitting the condensate to return back to the boiler.

If no steam space is provided and the system is 'closed' and simultaneously heated, its temperature will rise above that of boiling point (at atmospheric pressure) to a temperature equivalent to the pressure within the system – hence the terms *medium* or *high-temperature* hot water. Expansion of the water will obviously occur and it is the method of controlling this expansion, without reducing the pressure of the system, that produces three different types of system, namely:

(*a*) The method of using the steam space to maintain pressure on the closed system without drawing off steam.

(*b*) Imposing an artificial pressure on the surface of the water by inert gas such as nitrogen.

(*c*) Imposing a static head on the system from a high level feed and expansion tank.

Primary-Heat Generators can be designed to use any of the basic fuels or methods of firing ranging from solid fuels, liquid and gaseous fuels to electricity, or to combinations of some of these.

Primary-heat generation can also be obtained from 'waste heat'. This is the relatively low-grade heat available from industrial or commercial plant which requires to be operated far in excess of the normal temperature and pressure range applicable to comfort conditioning. Such 'waste heat' is obtainable from electricity generating plant and applied to district heating distribution-networks. The adoption of district heating has many advantages, particularly in the bulk use of fuel and its handling, together with the control of exhaust gases to the atmosphere.

PLATE III. Overhead Unit Heaters (down-draught) in a factory (*see page* 134).

Fuels for primary-heat generation

The history of fuels is a story of discovery and development which underlines the truth of the adage that necessity is the mother of invention. Perhaps it might even be said that the converse was equally true, since it has been the need to find means of realizing industrial projects which has led to the development of fuels, and, as the supply of one fuel has been seen to be limited, another has been developed or exploited.

Social standards, economics of production and transportability greatly influence the choice of fuels; 'house coal' or soft bituminous coal, for so long the main source of domestic heating, is declining, while the smokeless fuels derived from it are increasingly used. Similarly, the mechanical stokers, usually either a hopper feed over the fire or a moving-belt underfeed mechanism, are replacing or have replaced hand-firing, due to the economics of labour costs.

Oil fuels range from heavy crude oils of the consistency of tar to the light distillate oils. The heavy grades, being more viscous, have to be kept heated in the storage tank and run by heated pipes to the boiler burner. Even the lighter distillate oils can coagulate in sub-zero temperatures and may have to be kept warmed to obtain a flow in certain locations. Oils, like solid fuels, require storage areas, often at a premium in densely developed areas, and the storage must be of a form which is conducive to safety. The oil tanks must be installed to conform with local bye-laws and the fire officer's regulations. When oil tanks are contained within a bund wall, the capacity must not be less than the total oil-storage capacity plus 10 per cent. If the tanks are within a building, the oil-storage chamber is treated as a bund, access doors being set high up on the wall. The walls must have a high fire resistance and the chamber must be vented to the external air by fire-resisting vents (usually cast-iron soil pipe is used). Tanks can also be sunk into the ground; they are either of reinforced concrete and lined with ceramic or glass tile, or mild steel tanks, within a chamber or set in sand.

The cost of providing for the storage of solid or liquid fuels must be considered – a factor which does not apply to gas. Gas, supplied through mains, when available in adequate quantity, has much to commend its use; however, the supply of the fuel is outside the individual's control, which is not the case with the other two fuels. Gas is widely used in industry for process purposes, and this can often be an important influence on the selection of the fuel, since an even better tariff rate may result from increased use.

A number of industrial boilers are adaptable for both oil and gas firing, and this gives the owner an extra degree of security against failure of one supply or the other.

If there is a perfect fuel, this could be said to be electricity. However, like all things tending to perfection, it is relatively costly; but it certainly has applications, and its further development for heating over the next decade or so is inevitable.

One particular development of electricity, and perhaps the most interesting technically, is that used by the heat pump, in which low-grade heat – extracted from a river, ice rink or cold store – is 'upgraded' and used to provide heating. Current experience ranges from domestic installations to large civic/commercial enterprises, but it is still relatively expensive in capital investment for primary plant.

Some aspects of thermostatic control

There must be some controlling element between the fuel, primary-heat generation and secondary-heat transfer so as to establish the desired balance between the internal environment and the external climate. This is achieved through the Thermostatic Controls.

They are the link between primary-heat generation and secondary-heat transfer. The aim is to 'anticipate' the thermal requirements resulting from the external conditions and to 're-adjust' by interpretation from 'feed-back' of the 'actual' results experienced within the building.

Broadly, the controls fall into the following groups:

1. The control of the fuel in the quantity to meet the demand of the heat generator.
2. The control of the rate of dissipation of heat from the heat generator to meet the demands of the secondary-heat transfer-plant as a whole.
3. The 'element' or 'zonal' control of the secondary-heat transfer-plant within the space to meet the fluctuating demands of heat loss or heat gain locally.

The degree of control, in terms of the overall efficiency of the installation, is dependent upon the level of comfort conditions required, whether or not the thermal characteristics of the installation match the thermal characteristics of the building and its function, the type of system and the fuel used.

Associated with the thermostatic controls are the safety controls; these are the limiting devices which normally only function in the first stage when the link between items (2) and (3) have broken down, and finally attempt to render the whole installation 'safe' if there is a further breakdown in the link between items (2) and (1).

Maintenance and safety

The maintenance of boiler plant, particularly of the complex automatic plants, is a matter for skilled attention. Most large and complex installations are supervised by maintenance staff under the direction of a works engineer. In small installations, however, it is usual to employ the services of firms who maintain plant on a contractual basis, leaving day-to-day supervision to the care of a general houseman. Similar arrangements are available for domestic installations – normally, through the fuel suppliers.

Life-span of Plant

Buildings are erected with an approximate life-span in mind. Some buildings, such as cathedrals, may be expected, or intended, to last indefinitely, while others are short-term. Generally speaking, the depreciation of buildings in financial terms may be regarded as spreading over a fifty to seventy-five-year period, though the tendency appears to be to consider shorter terms as the true economic life of a structure, measured more in terms of its functional viability and comfort standards than of its structural stability or architectural durability. It is, however, unlikely that any structure which takes twenty-five years to pay for (on mortgage basis) will be dismantled within a year or two of its purchase-completion, unless some system of trade-in, such as applies to other consumer durables, has been established. Thus, it is more probable that, though many commercial and some industrial buildings, whose earning power can be measured in terms of output per square root, may be more rapidly depreciated than hitherto (subject to a major shift in tax policy), most other buildings – particularly the domestic and the monumental municipal – will still be expected to function for fifty years and upwards.

Such durability is not unusual in construction and detailing, but a different standard must be applied to the service of buildings. The static and low stress nature of building shells puts small demands upon the capabilities of the materials used, while the constant movement and rotation, the flow and resistance to flow, i.e. the dynamic nature of servicing, inevitably puts a strain on the materials used in plant and servicing. Thus, the generally accepted life of a boiler may be around twenty years – perhaps shorter, if inefficiently used. Pipework can, if precautions are not taken, be rendered useless by furring and scale or acid attack in a much shorter time. Pumps, motorized valves, fans, etc., if not properly and regularly maintained, may have to be replaced within a year or two.

Scale and Furring

Boiler scale and the furring of pipes are similar. They arise from the hardness of the water used in the system, hardness being the calcium compounds. Some of the compounds are easily removed by boiling, while others require more detailed treatment to remove. In the boiler system, it is therefore necessary to try to use the same water over and over again so that the quantity of hardness precipitated out of the water is as small as possible. Due to evaporation loss, a certain amount of make-up water is inevitably needed, but otherwise the system should be such that no water is drawn off. With hot-water systems, this object is usually accomplished by using a water-to-water heat exchanger, where a coil or pipe, forming part of the 'primary' circuit, heats the water in a cylinder forming part of the 'secondary' circuit from which domestic hot water is drawn off. The relatively low temperature to which the secondary water is heated prevents undue furring of secondary pipes, while, since no water is drawn off primary circuits, only the initial hardness of the primary water is deposited in the primary circuit. The primary circuit includes the space-heating elements as well as the water/water heat exchanger (indirect cylinder or calorifier).

In soft-water areas, where it is possible to use direct systems, primary water is used for all purposes, the cylinder being only a tank in the circuit designed to increase the total quantity of water available for draw-off.

In steam systems where process steam is required for industrial purposes, it is usual to draw the steam direct off the steam circuits. Clearly, it is advisable to keep such irrecoverable steam to minimum quantities, but, if such losses are unavoidable, the engineer has to consider water treatment to ensure that hardness is reduced before make-up water enters the system.

Water treatment can be accomplished in several ways. Distilling water by condenser is a method which is suitable for the laboratory and for special processes, but it is too expensive for large scale and wasteful use. Base exchange, using zeolite crystals, is a common domestic water-softener method. Zeolite (sodium alumino-silicate) absorbs calcium from hardness compounds dissolved in hard water, exchanging sodium in its place. The sodium compounds do not form from the heavy and hard scale ('fur') caused by the calcium compounds. Re-charging of the zeolite is easily accomplished by passing salt water through the crystals which then revert to their original chemical form by exchanging the absorbed calcium for the sodium in the salt (sodium chloride). The resultant calcium chloride is carried in the charging water to the drains.

Caustic soda (sodium hydroxide) can also be used for removing temporary hardness from water. It is not a suitable commodity for domestic use, and carries other disadvantages in industrial use. Washing soda (sodium carbonate) removes both temporary and permanent hardness.

The most easily maintained system of water-softening is by irradiation with special lamps which cause the hardness to precipitate in a receiving well. This system has the advantage that a constant flow can be maintained without the fluctuations in hardness arising from the dosage systems required for chemical precipitation and conversion. The other systems also leave sodium compounds dissolved in water.

Accessibility

Nevertheless, even with all precautions and maintenance, environmental-control systems have a limited life and may have to be completely renewed every decade or so. They should therefore be accessible, and this requirement can bring a basic conflict between the architect and the mechanical-services engineer. The balance of concealment and accessibility is a fine one and should be handled with understanding, so as to achieve a result which is pleasing and efficient as well as being accessible for plant replacement.

To be replaceable does not imply that every part of the system should unscrew and pull out of clips. The task of replacement is bound to be a pretty far-reaching and disturbing exercise for all concerned and, if a bit of flame-cutting or dismantling of pipe casings is involved, it need not be regarded as fatally bad design.

Movement

It is equally important that the secondary effects of a heating installation on building maintenance should be appreciated and provided for, when possible, in the detailing of buildings. The passing of hot media through pipes and ducts causes them to expand both diametrically and longitudinally, so it is necessary to allow room for this movement in installation. Pipes, for example, are sleeved where they pass through walls so as to allow them to slide without thrusting against the structure and without buckling. Similarly, the heating elements expand when heated and require to be fixed in such a way as to permit movement. Pipe supports should be flexible or fitted with rollers, to permit this movement to take place. With high temperature installations, more precautions have to be taken to integrate them into the building: for example, long straight runs require expansion bellows and detailed design of pipework to accommodate the thermal movement to be expected in the system.

Casings

The heat given off by pipes and ducts creates local air currents which carry dust and cause staining. Thus, it is necessary to seal all casings to pipes and convectors to reduce the staining to a minimum. The casings themselves must be protected from the heat by insulation, otherwise they may warp or distort. Timber pipe casings are very subject to warping unless heavily insulated, and it used to be common to cut ventilation holes in them to induce convection currents, thus dissipating some of the heat. This, however, results in dust staining and, consequently, natural wood casings are less popular than particle board or metal casings. Particle boards, such as chipboard, have better qualities owing to their homogeneity and lack of grain. Their warping can be forecast and prevented, whereas natural wood tends to twist and curve in both planes. Metal casings can be neat and durable but are not as convenient in use for irregular runs as the types previously mentioned. The casings of convectors, including continuous convectors, are almost always made of pressed-metal units, with special filler lengths to make up non-modular dimensions of continuous runs. They are always provided with dust-sealing strips where they touch walls.

The effect of heat on timbers has been noted earlier; thus, all timber in close contact with heating installations should be carefully considered, both from the point of view of warping and of desiccation of the timber. Wood relies for its strength upon the moisture content, and over-dried timber becomes more brittle as well as being more liable to catching fire.

The *Building Regulations* control the proximity of timber to flues and fireplaces, but the architect should also be conscious of the fire hazards attaching to other sections of heating installations.

Building for safety

Fire

Previously, we mentioned the built-in safety measures for boiler plant, but the architect must also create safe housing for this plant, bearing in mind that it is inherently more liable to fire than most other parts of buildings. The safety of personnel is the first consideration, and it is therefore essential to provide adequate means of escape. Usually, two exits at opposite ends of the space are required, the doors being self-closing doors opening in the direction of escape, and being of at least a half-hour standard of fire resistance (see B.S. 459, Part 3).

Next, to contain the fire, the walls of the boiler-house should be incombustible and of high fire-resistance (usually four hours: see *Schedules* to *Building Regulations*). The fire brigades generally prefer to be able to get smoke away from buildings rather than to bottle up the fire. This enables the firemen to get into the space and tackle the blaze without being choked and blinded, so it is common to provide smoke outlets – sometimes automatic roof vents, which operate when an excessive temperature is reached. To prevent the spreading of oil in boiler installations, a curb of concrete forms a threshold at all door openings, and pipe trenches in floors are sealed at the wall line.

To help to fight the fire, it is now common practice when

using oil fuel, to provide a foam inlet to which the fire brigade can connect their foam unit and pump the boiler-house full of a blanketing foam which kills the fire by eliminating oxygen.

Where an oil storage chamber adjoins the boiler room, the access doors are of the fire-resisting, self-closing steel type, providing at least two hours' resistance to the passage of flame. Boiler attendants frequently fail to appreciate the importance of the correct functioning of all the doors in a boiler-house and notices are usually desirable, informing staff of the necessity to keep all fire doors shut and exitways clear.

Another hazard of boiler-room plant can be the 'blow-back', resulting in a fire. Once established, this fire may find an easier exit from the combustion chamber than the flue and may persist for some time. Therefore, for the safety of the attendant, the escape doors mentioned earlier must be provided in such positions that no matter where the man stands he can escape without having to pass in front of the burner. Oil-fired installations always include a fusible link or other sensing device which cuts off the supply of fuel in the event of a blow-back.

Scalds and Burns

In steam or high-temperature hot-water installations, there is a risk to persons in the possibility of steam blowing from joints or valves. The safety valve discharge should be designed to blow-off without danger to persons. Careful maintenance is the only way to reduce the risk, together with sensible operation of the plant within its designed limits.

Steam and high-temperature hot-water pipes and fittings are very hot – hot enough to cause severe burns if unprotected. Therefore, all possible surfaces are insulated to conserve heat and to protect personnel. A well-designed pipe layout, with the lagging painted in the appropriate B.S. standard identification colours, can produce a lively and stimulating interior, and a good standard of durable and washable finishes to the structure can not only enhance the initial appearance but also encourage good maintenance and tidiness in the boiler-house.

Flues

When flues are required, the problem is now a matter of conforming to the *Clean Air Act* and of specific design at the flue outlet to ensure adequate discharge velocity is achieved. It is also good practice to provide separate flues for each boiler rather than one large flue. The flue height at a domestic level often depends to some extent on the nature of surrounding buildings, trees, etc., which can cause peculiarities in wind direction.

In most chimney design, it is necessary to protect the structure from undue thermal stress and to maintain flue-gas temperatures above their dew-point. Thus, for about two-thirds or more of the height, a lining of refractory (heat-resisting) bricks is formed. Such bricks are diatomaceous, being soft and porous insulating brick, and subject to deterioration if subjected to the elements. Linings are often separated from the main structural stack by a cavity, but are usually then stabilized by bricks projected from the structural stack to give support.

Metal flues, either self-supporting or guyed, form a very slender pencil-shaped stack and can be architecturally effective in the correct circumstances. They are made of mild steel sections with flanged joints and are insulated and protected with aluminium cladding.

In principle, the best cross-section for a flue is the circle, which permits the spiral rotation of rising gases to occur without the sharp changes caused by corners in a square flue. The cylinder flue is not always convenient to build, however, and rectangular flues may even be dictated by circumstances. The aspect ratio of such flues should never exceed 2 : 1.

Domestic flues are required under the new building regulations to be lined with circular flue liners, always kinked in their run in order to prevent direct rain on the fire and to reduce the effects of down-draughts. Such flues should slope at an angle of 45 degrees or more. The steeper the slope, the more easy is the passage of flue gases.

The burning of all hydrocarbon fuels produces solid wastes – soot, in the case of coal and coke; lamp black from oils; and chemicals which combine with air and water to form solid or liquid compounds, in the case of gas. With gas, it is particularly important to realize the nature of the compounds formed. SO_3 and water are two of the products of gas and oil combustion. SO_3 under certain circumstances can form a hard deposit at the base of gas flues and around the firing chamber, but it also combines with water to form H_2SO_4 (sulphuric acid) in a dilute form. Corrosion and the breakdown of surfaces follow attack by the acid – hence the staining which occurs near flueless fittings, and the necessity to use flue liners for gas or oil installations, and to provide adequate ventilation.

Primary-heat generation (plant)

The foregoing generalized discussion of primary-heat generation has touched on the wider aspects of heat systems together with fuels and certain building and maintenance features. In order more fully to comprehend the actual plant required in the various systems outlined, it is necessary to consider the basic layout patterns of plant items and their relationship to one another and the building itself.

Primary Elements of the Installation

A. Steam

This requires a boiler or steam generator, condensate collecting tank, boiler-feed pumps, water-treatment plant and raw-water make-up tank. All this equipment can be located locally in the boiler room.

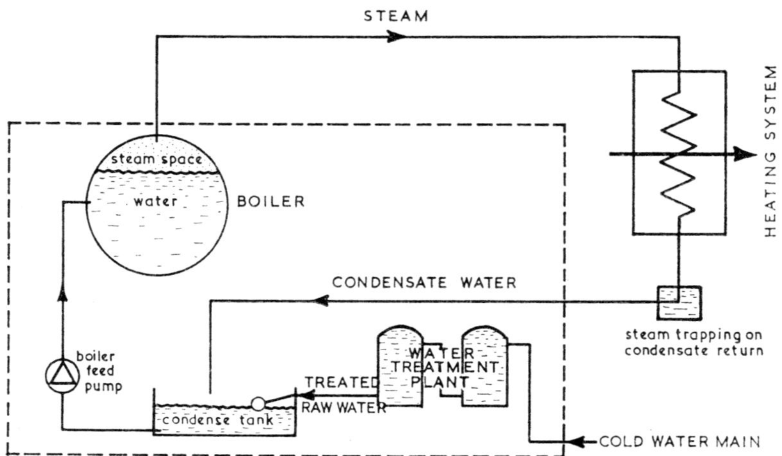

FIG. 13.1. Typical Steam Installation. (Applicable to industrial use, but not applicable to a roof-top plant room.)

B. *Hot Water (see Figs.* 13.2 *and* 13.3)

(i) L.P.H.W. requires a heat exchanger (i.e. boiler or calorifier) and *Open-Head Tank* (situated above the highest point of the system), and a pump (if not a gravity system).

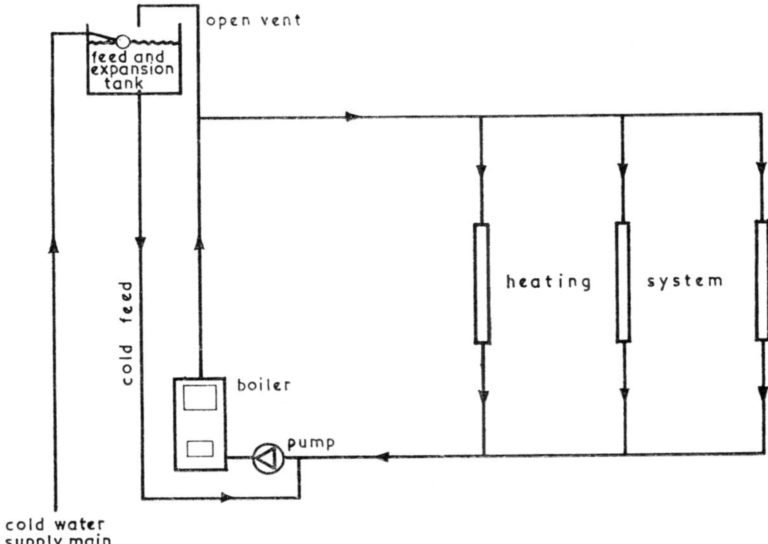

FIG. 13.2. Typical Low-Pressure Hot-Water Installation. (Applicable to domestic and medium-sized commercial jobs.)

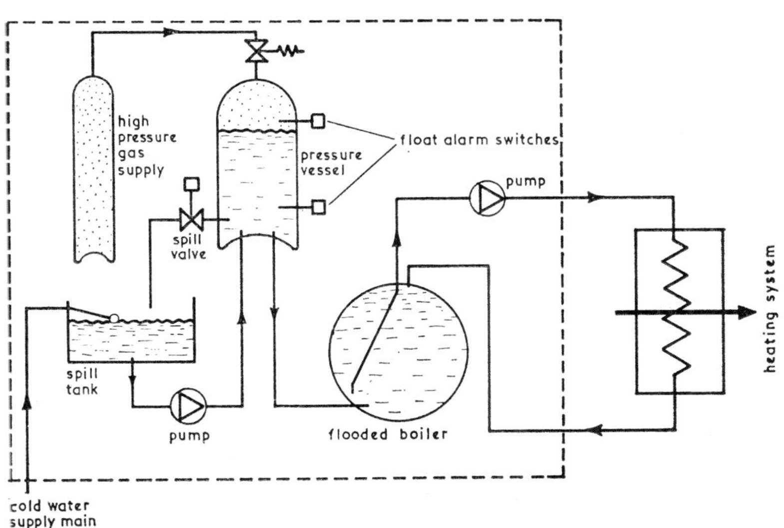

FIG. 13.3. Typical Medium or High-Pressure Hot-Water Installation. (Used in industrial and large commercial jobs and for district heating).

(ii) H. and M.P.H.W. requires a boiler and a closed system working under pressure (normally artificially imposed by use of nitrogen), a feed pump and spill tank and circulating pump located within the plant room.

C. *Gas*

Gas can be used either (*a*) directly, or (*b*) indirectly:

(*a*) If used directly, it requires provision of a gas meter and distribution pipework to the local heaters.

(*b*) If used indirectly, it requires a gas meter *and* heat-exchanger plant to produce mediums A or B.

D. *Electricity*

Electricity can be used either (*a*) directly, or (*b*) indirectly:

(*a*) If used directly, it requires only transformers and switchgear.

(*b*) If used indirectly, it requires transformers and switchgear *and* heat-exchanger plant to produce secondary-heating medium (A) or (B).

E. *District Heating*

When a number of buildings are to be built as a single development, the individual heat-generating units can be dispensed with and each building served from one centralized boiler plant using steam, M. or H.P.H.W. to provide a thermal

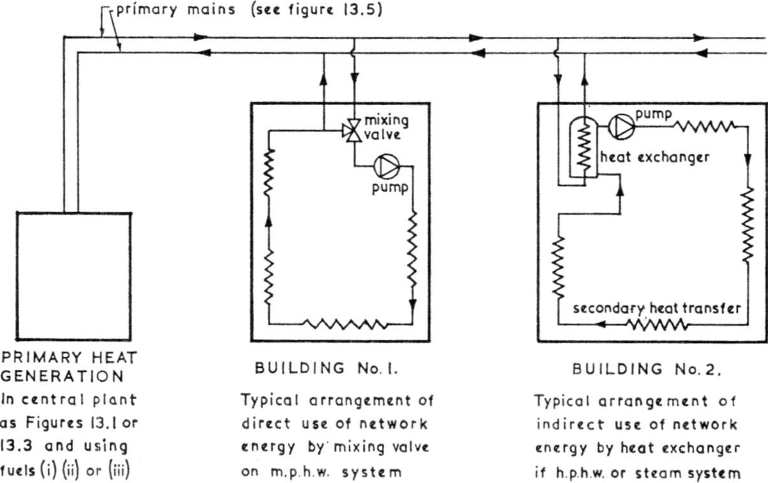

FIG. 13.4. Typical District-Heating System.

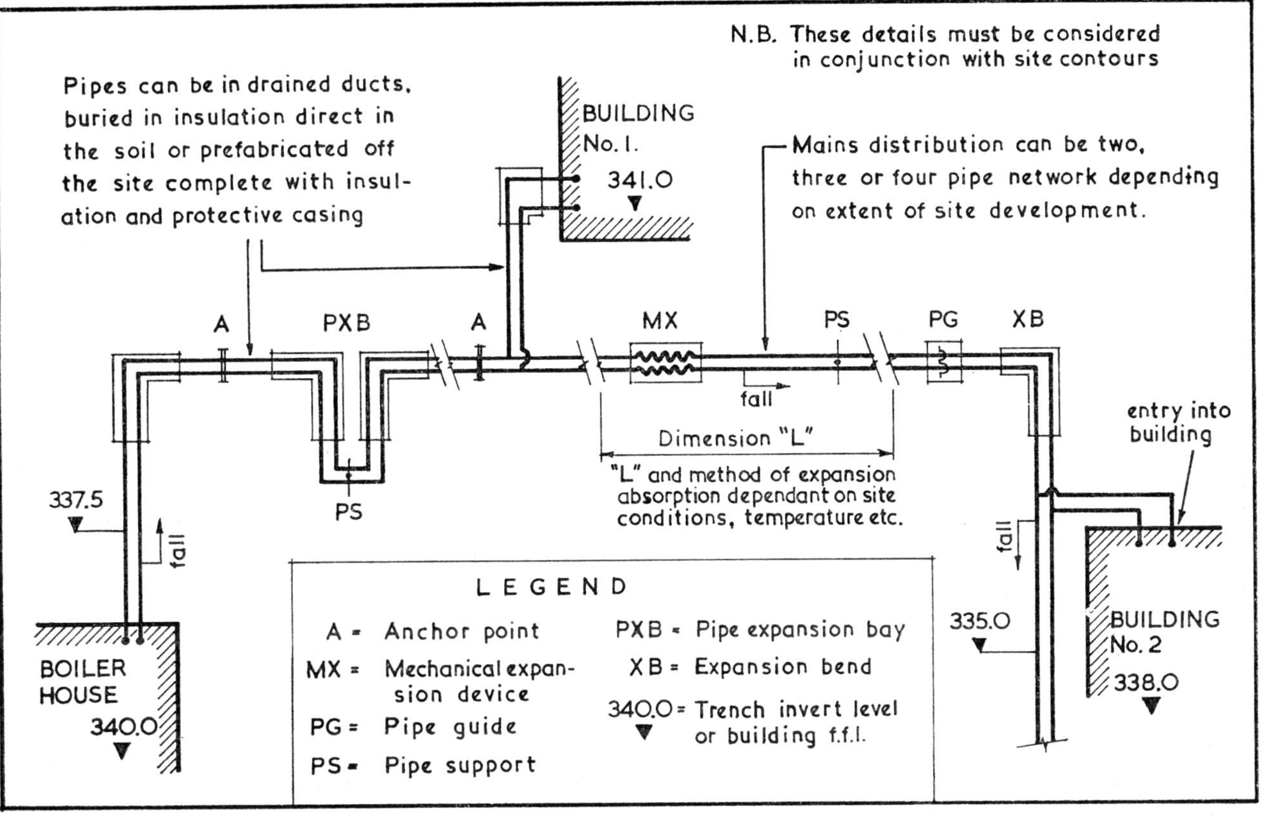

FIG. 13.5. Typical Thermal Network Distribution.

distribution network or district-heating scheme. (See Figs. 13.4 and 13.5.)

The provision of domestic hot-water supply is outside the scope of this book, but, like so many other services, needs to be considered as an integral part of the overall service require-ments. Therefore, if some of the principles, governing which type of H.W.S. system can be linked with a particular type of heat system, are noted, the details can be investigated more fully in standard technical books.

At this stage, digression from the consideration of heating is necessary only in order to assess, in outline, the effect of this secondary requirement on overall plant size and its location.

Domestic hot-water supply

This can be supplied as an integral part of the heating instal-lation or individually by gas or electricity.

There are three methods of providing domestic hot-water supply:

(1)	(2)	(3)
Instantaneous	*Separate System*	*Combined System*
By gas or electric water heaters	Using storage cylinder and boiler separate from the heating system	Using storage cylinder as an integral part of the heating system

1. *Instantaneous*

This method does not require allocation of space within the central plant, but does require adequate space (effective working headroom, in the case of the electric self-contained tank and storage-type heater) adjacent to the location it is serving.

Supply pipework generally is single-supply (dead-leg) without a secondary return.

2. *Separate System* (*see Fig.* 13.6)

The hot-water boiler and direct cylinders can either be separ-ate or combined units, but in no circumstances must they be interlinked with the central-heating system. They require plant space which can be within the central plan area or, if more prac-tical, remote from this and adjacent to the location they are required to serve.

Pipework can either be single-supply (dead-leg) or with a secondary return.

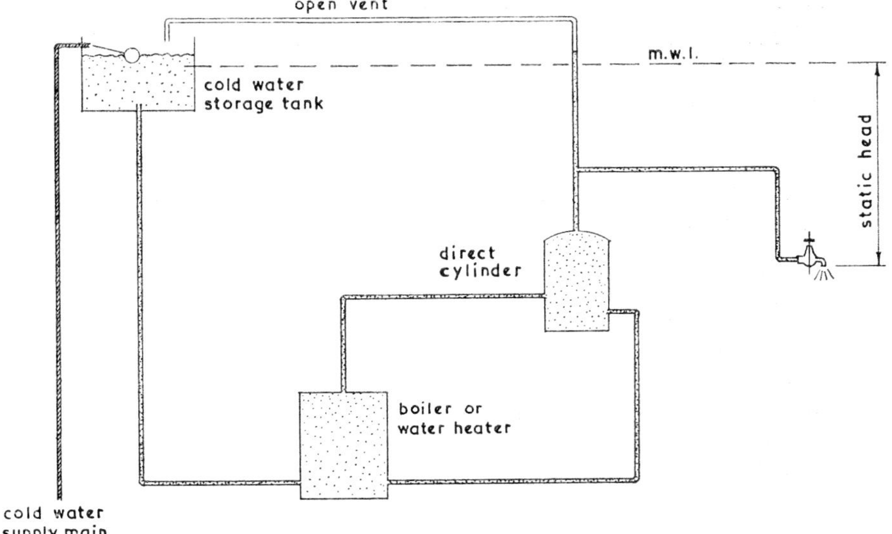

FIG. 13.6.   Separate (Direct) System of Hot-Water Supply.

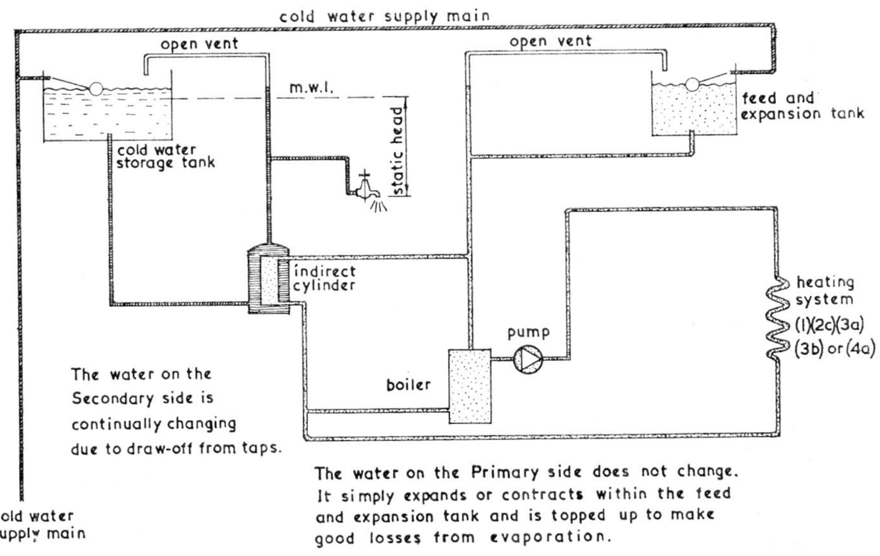

FIG. 13.7.   Combined (Indirect) System of Hot-Water Supply.

3. *Combined System* (*see Fig.* 13.7)

This method can be combined as an integral part of the heating plant, and whilst it is normal practice to install bulk hot-water indirect storage-cylinders within the heating-plant area, this is not an essential feature, and these cylinders can be located adjacent to the areas they require to serve. In large thermal network distributions, it is in fact the practice to locate these calorifiers where required within the individual buildings.

General

With both the separate and combined systems, bulk cold-water storage is necessary to provide the static head for distribution, whilst system (1) can be operated direct from mains services, at main pressure; however, the number of such units accepted on the mains supply would be governed by the local water authority.

On the Continent, the individual bulk cold-water storage tank is often omitted when using the combined system, and the incoming cold-water main is pressure-controlled into the indirect cylinders. This type of indirect cylinder is sometimes referred to as a *Bubble-Top* cylinder and is only permitted in this country under certain conditions.

In the separate (direct) system, see Fig. 13.6, the water is *common* to the *tank*, *boiler* and *cylinder* (limited to small domestic systems) and, in consequence, is continuously being replaced throughout the system, which can result in furring-up as the chemicals within the water are deposited out. For this reason, it is not good practice to include radiators as their relatively small water-ways soon become choked in hard-water areas.

The combined system is tending to be universally adopted, particularly with the increase in domestic heating, and is applicable to all installations requiring H.W.S. in quantities exceeding domestic requirements.

When integrated with a district thermal-network system, it is essential to use the combined system, at the same time eliminating the primary boiler plant. This is shown in a simplified form in Fig. 13.8.

Local *electric immersion heaters* or *gas circulators* for provision of domestic hot-water supply can be used as an integral part of the indirect cylinder, making it into a 'direct system' when the primary heating is shut off. With direct electric- or gas-

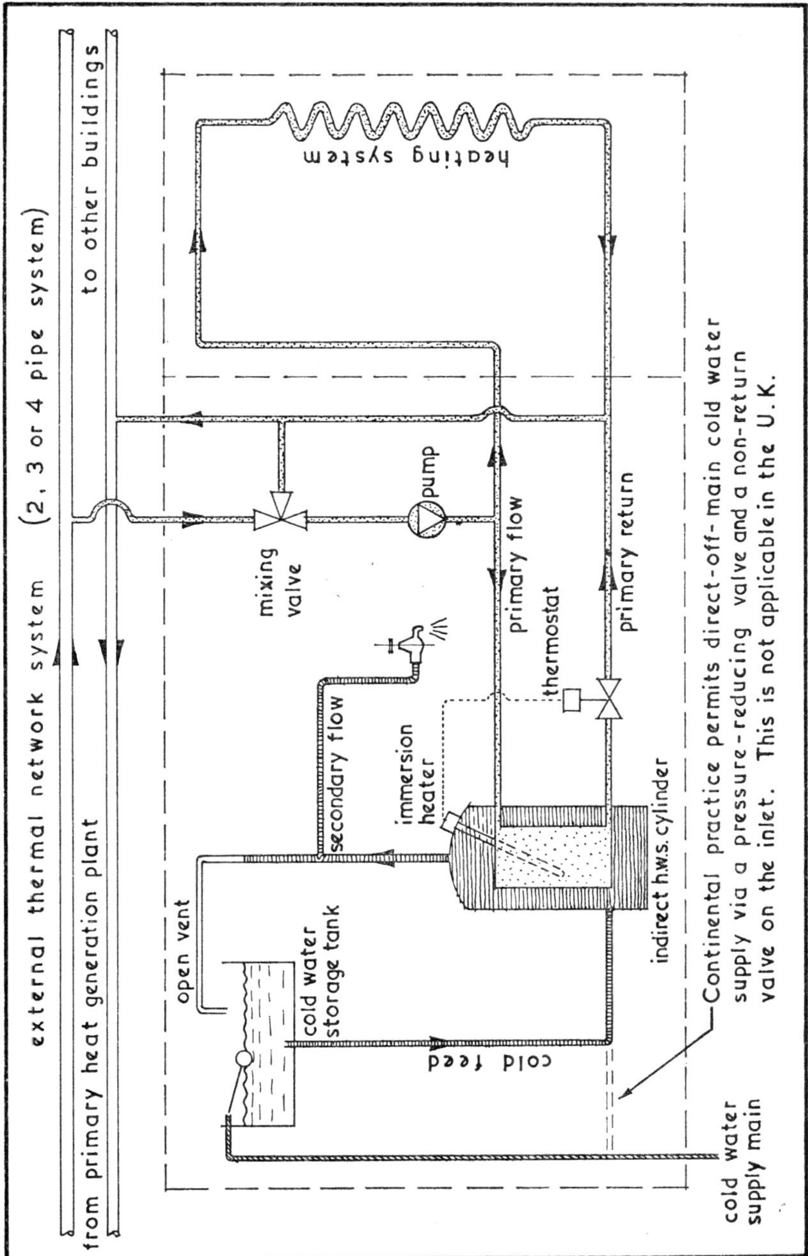

Fig. 13.8. Typical h.w.s. Service from Thermal Network.

heated storage-type H.W.S. cylinders the same arrangement as Fig. 13.6 is applicable.

Thermal control

By the words *Thermal Control*, the link between primary-heat generation and secondary-heat transfer is implied.

Having established that, within a building, it is always necessary to think in terms of thermal zones, these can be small isolated locations with individual control, or larger areas or whole buildings, dictated by the orientation or geographical location of the building. The technique or methods of providing controlled environment are numerous and fairly complex and can involve a detailed study in themselves. However, it is necessary for the student architect to be broadly aware of some of the general terminology and the extent to which some methods of control can be used.

The control requirements can be considered to cover three stages:

(i) Control and regulation of primary fuel.

(ii) Thermostatic control of the Primary-Heat Generator to obtain the design-operating conditions *within safe limits*.

(iii) Thermostatic control of the secondary-heat transfer to obtain the *balance* between external climate and the internal environment, so as to maintain acceptable comfort level for the *occupants*.

Each of these is obviously interrelated and interdependent, but the 'degree' of control of each section is influenced by different factors, namely:

(i) The type of fuel and its application will be more closely controlled the more expensive it is in primary cost.

(For example, the practice of purchasing electricity, metered and controlled, 'off-peak' to obtain the most economical rate of intake, compared to the use of solid fuel in which a minimum firing rate must be maintained even though, in theory, no heat is required and in consequence no fuel need be used.)

(ii) Having decided on the limits of design for primary-heat generation, the controls are designed to operate the plant within these limits. If these limits are exceeded, for any reason, the controls should shut-down the plant and cut-off the fuel supply.

(iii) The 'degree' of control or tolerance within the occupied
space is dependent upon:

    (*a*) The necessity to maintain the space at a controlled
level independent of external climate and internal
heat gains, or, within reason, letting the internal
conditions fluctuate with variations in these two
factors.

    (*b*) Statutory or other legal requirements laying down
minimum acceptable conditions for a particular
function.

    (*c*) The maintenance of internal conditions at a certain
level, which would otherwise cause deterioration of
stock and fixtures or the building structure.

Types of thermal controls

A. *Mechanically Actuated, Direct-Acting Controls*

In which the controlling valve is actuated by the expansion or
contraction of fluid vapour in direct link with a phial or element
by capillary tubing.

Control Application	*Remarks*
ELEMENT CONTROL Local heating elements, radiators, convectors, etc.	These can either be pre-set or adjusted manually by individuals; such as thermostatic radiator valves.
ZONAL CONTROL 3-way mixing valve	These are located in the heating main and operate from an external climatic condition and vary the mixed-flow temperature to the zone.
PRIMARY CONTROL Direct acting on/off or modulating	These are principally used to control the primary water to calorifiers (on/off) or heat exchangers (modulating).

B. *Direct: Electrically Operated*

The thermostat or sensing element acts either directly or
through a contactor which stops or starts the pulling motor on
valves, the motor driving the fan or pump, or on solenoids
operating valves, dampers, etc.

Control Application	*Remarks*
ELEMENT CONTROL Individual heating elements	The room thermostat opens and closes the control valve, or stops and starts a fan motor.
ZONAL CONTROL Groups of heating/units	Two or three thermostats can be used as an averaging control in which case any two calling for or rejecting heat will operate the valve.
3-way mixing valve or 2-way modulating valve	These are usually modulating valves and respond to an internal sensing element and an external pilot operating through a central controller giving the required flow temperature.
3-way diverting valve	On response from a room, duct or immersion thermostat the primary water is diverted past the heater batteries or calorifiers thereby reducing the heater output.
Dampers or louvres on fresh-air inlets or by-pass ducts	Used in conjunction with plenum ventilation plant and operated from duct thermostats, room thermostats or other sensing element.
Humidifier	This is controlled from a Humidistat which responds to the moisture contained in the rooms, or a thermostat giving fixed dewpoint condition.

The electrical-control system is completely flexible and independent of the location of the various components within the building.

C. *Pneumatic/Mechanically Operated*

Pneumatic: The controlling valve is actuated by compressed air providing the motive force which, in turn, is actuated from a local direct-acting thermostat reacting to room conditions.

Electro-Pneumatic (Mechanical): A combination of remote control by electric sensing-elements and pressure-controllers can be used to actuate the control valves or dampers.

This control system has applications to the same extent as the fully electrically operated system, but requires the provision of capillary tubing and compressed air to actuate the components.

The controls described above link the Primary-Heat Generation with the Secondary-Heat Transfer System and, between them, control the overall efficiency of the installation in two ways:

(a) the thermal efficiency of the plant (the heat energy usefully transmitted expressed as a percentage of the total heat energy of the fuel supplied);

(b) the efficiency with which the internal climate can be controlled to provide comfort conditions.

THIS CONCLUDES the consideration of thermal plant in respect of heating and ventilation, the ascertainment of acceptable comfort levels within the limitations of this plant, and its effects and requirements as an integrally functioning item of a building complex.

Control of the 'degree' of comfort obtainable cannot be extended much beyond the heating season, and, in fact, we will see in the following chapter that, in certain circumstances, internal heat gains are a problem all the year round, irrespective of external climatic conditions. In general, however, it is the seasonal thermal change which imposes the problem of providing comfort for a condition of heat loss in the winter and heat gain in the summer.

14: Providing a Fully-Controlled Environment

General

THE ELEMENTARY PHYSICS AND PHENOMENA so far discussed for heating are equally applicable to cooling. Cooling is the extraction of heat. Therefore, cooling can only be carried out by supplying a cold medium to the building spaces in order to pick up unwanted heat, or by inducing a high rate of evaporation from the fabric of the space. The second method may work very well for cooling a bottle of milk in a porous clay pot dipped in water, but it goes without saying that dealing with a building in this manner would tax the architect's abilities as well as those of the design engineer. Hence, the solution of cooling problems is normally by means of a refrigerant medium.

Central cooling

Cooling the actual fabric of a building is undertaken only in cold stores, etc., problems of circulating chilled liquids through walls being much more difficult than circulating hot water, largely because of the effects of subsequent condensation; if the dewpoint and R.H. are kept low enough this method could be adopted. Generally, a building is treated much as a domestic refrigerator using a localized centre of heat extraction (the evaporator) and letting air movement (convection) do the rest.

Thus, it is similar to the heating problems in that two choices are available, taking either a cheap chilled medium (water) to a number of heat-absorption units, or passing the air through a central point where the heat is extracted from the air which is then passed (secondary-heat transfer) into the conditioned space to pick up heat before recirculation to the central plant.

Chilled water circulation by lagged pipes, rather like a heating distribution, is commonly used in conjunction with primary pre-conditioned air, or all-air systems for larger installations, but a common method of local cooling is the use of 'air-conditioning' units. These units incorporate their own refrigeration system and draw room air over a cooling coil, mixed with some fresh air, and recirculate it back into the room. The heat is extracted from the refrigerant *via* the condenser by drawing outside air over it, and it is then blown to the external atmosphere. The units usually require a fresh air make-up, and can often be reversible so as to provide heating in cold weather from the same unit. In their simplest form they are mounted in the window or external wall.

The centralized cooling system using blown chilled air is an extension of the warmed air-heating ventilation system requiring a heating coil transferring heat into the airstream and a cooling coil extracting heat from the airstream, dependent upon requirements within the space. A properly devised duct-distribution system can be used for both, with the additional requirement of a recirculation duct from conditioned space to central plant. There are three basic types:

1. Pre-treated all-air systems with supply and recirculation ducting.
2. Chilled water to terminal units with localized fresh-air inlets for ventilation rate.
3. Combination of pre-treated primary air and terminal units requiring chilled water.

Remembering, however, that cooling of the air introduces condensation under normal conditions, all must be provided with drains to allow this condensation to be removed.

The cooled air, on being discharged into a room, is heavier than the warm existing air; thus, if it is injected at low level it will lie like a cold puddle around the feet. Hence, the normal place to introduce cold air into a room is at a higher level, and in general the entries should be numerous and well spread so as to produce an even passage of cold air through the space picking up heat.

Humidity changes

The cooling of air below its dewpoint produces condensation which must be drained away. Thus, the cooled air, though practically saturated at the low temperature, will have a reduced humidity when it picks up heat by circulation through the conditioned space. The result of a continued cycle of cooling without adding moisture is therefore the progressive lowering of the relative humidity of the atmosphere within the conditioned space. In actual fact, only the air directly in contact with the cooling coil should reach dewpoint, and the condensation takes place on the coil surfaces rather than on the structure. The extent of dehumidification of the air may reach a point of balance when the fresh-air make-up contributes as much moisture as may be lost by the air. However, the specific humidity may still be low enough for the room atmosphere to be dry at the temperature required, and it becomes necessary to increase the humidity of the air. This is done by using spray or capillary air washers, or direct injection of moisture into the air stream, either mechanically or by evaporation.

Dehumidification also arises in winter with low outside air temperatures. The air, being saturated and having a high R.H. at this outside temperature, has a very low R.H. once it is heated to the room condition. It is therefore often necessary, during the dry cold weather, to increase the humidity to avoid uncomfortable conditions.

Cost, 1967

To provide cooling is considerably more expensive in capital expenditure and running costs than to provide heating. The installation costs to supply heating might range from £12 to £17 per kW (£3 10s to £5 per 1000 Btu/hr), whereas the equivalent costs for providing both cooling and heating might range from £60 to £120 per kW (£17 to £34 per 1000 Btu/hr). In consequence, there has to be some decisive factor demanding the use of cooling before this element can be added to the basic thermal-environmental cycle required for comfort. The primary consideration must be 'can comfort conditions be achieved and maintained for a given set of circumstances within an occupied space without the use of artificial cooling?'

The totally enclosed zone

Referring therefore to the equations of thermal exchanges (pages 58, 59), it will be seen that the internal sources of heat gains during winter are normally offset by external heat losses. For example, under normal circumstances, in an area which has exposed external surfaces being heated, the effects of solar gains through the glass, or the increase in lighting or machinery load, should be counterbalanced by the thermostatic controls for the heating system as indicated in the expression for balanced thermal environment.

Effects

If an equivalent area to the above is now totally enclosed within the building structure, the equation for thermal comfort begins to become unequal (neglecting the necessity for ventilation), the heat gains are maintained – and probably increased with a higher lighting intensity – but at the same time the means of heat loss have been eliminated. In consequence, if the heat gains continue without a source of heat loss, then there is an increase of space temperature. Up to a point, the body adjusts its rate of heat dissipation by over-emphasis of the natural reaction of perspiration with the consequent evaporation of the moisture from the body, referred to previously as *evaporative cooling*. The rise in temperature within this space is occurring *despite* the conditions existing in zones around the area which, due to exposed surfaces and consequent heat loss, might well be required to be *heated* to maintain comfort conditions. Simultaneously with the rise of internal temperature, with the additional moisture added to the air, the humidity will increase as temperature increases, and conditions outside the 'thermal index' of comfort conditions can result. Add to these factors the complete lack of fresh air, to provide oxygen and to dilute the vitiated air within the area, and it will be obvious that, at the worst, conditions would become increasingly adverse for human comfort, and at the best would produce sluggishness and a preoccupation with the individual's discomfort to the exclusion of anything else.

Treatment

In this case, we have the conditions which could prevail in an internal room of a building irrespective of external conditions. If we now provide adequate ventilation, what is the effect going to be? Introducing outside air by ducts at say 1°C (30°F) would certainly alleviate the heat-gain problem, but, as it is normally not conducive to comfort to bring cooled air into a room at more than 10°C (20°F) below room temperature, this in itself is not a practical proposition. Introducing the same amount of air during summer at say 25°C (80°F) would only aggravate the problem of an already high internal room temperature. It can be seen, therefore, that in order to meet these requirements the incoming air must be heated in winter and cooled in summer and, in consequence, the humidity must be controlled also. Within limits it is possible, by mixing the room air with the incoming air required for ventilation, to achieve a more acceptable condition. Whether this is sufficient depends on psychrometric analysis of the respective components enumerated in the thermal equation and their relationship when plotted on a psychrometric chart.

Other zones

Considering now a location with external surfaces — in particular, during the summer, — reference to the thermal equation indicates the various sources of heat gain which can and do occur. If a building is subjected to these simultaneous heat gains (due to its geographical location, orientation and exposure), then, depending upon its construction (i.e. ratio of glass to wall, degree of thermal insulation, etc.) and the heat gains from lighting, personnel or machinery, the internal temperatures will increase to such an extent that discomfort will be experienced by the occupants.

In hot or sub-tropical climates it is common practice to provide varying degrees of cooling. In more temperate climates such as that experienced in the U.K., this is by no means accepted practice, mainly due to the cost and the fact that, whilst these heat gains do occur, their full effect is offset by the relatively low ambient air temperature (on average 23°C (74°F)) and by the simple expedient of opening large amounts of window and the provision of sun-blinds. But if the facility of opening windows is denied due to other reasons — such as the attempt to eliminate noise and dirt, or the simple inconvenience of opening a window sixty metres up, with its attendant difficulties due to higher wind velocities — then the problem of dissipating the heat produced within the building has to be faced and tackled.

It should by now be evident that the necessity for cooling can be independent of the ambient air temperature when either solar gains are high or relative heat loss is low. Failure to provide cooling can produce long-term results which progressively get worse. Figure 9.7 (see page 100) indicates internal and external temperatures and humidities of a building in which opening windows were eliminated and only mechanical ventilation was provided. For part of the summer the resulting conditions were acceptable, but, when the full effects of solar and internal gains occurred, the resulting conditions became intolerable. It is interesting to note that, without exception, the maximum internal temperature peaked after the occupants left, due to the thermal inertia of the building, but then began to drop as the building came into shade and/or the lower external temperatures occurred. This example indicates the limitations of simple mechanical ventilation when coping with the full effects of heat gains and the thermal inertia of the structure.

Conditioning

When we talk of 'heating' or 'cooling' in relation to the internal thermal environment of a building, it implies 'conditioning' of the atmosphere to a greater or lesser degree. Thus, we have discussed in previous chapters the simpler forms of 'conditioning' the atmosphere from basic heating and natural ventilation up to the stage of using mechanical ventilation and mixing air within the internal space, with air at external atmospheric condition. With the addition of 'cooling' to the relatively simple thermal-environment cycle so far discussed, the 'complete conditioning' of the internal atmosphere can be achieved. There are numerous ways of implementing this, but, primarily, the requirements for human occupation involve air-handling equipment. Depending upon the degree of conditioning required and the relationship of the following three factors,

(*a*) Sensible Heat Gains,

(*b*) Latent Heat Gains,

(*c*) Ventilation Rate or air change,

the full range of installations from an all-air system to independent conditioning units may be used. The choice of the system that can be adopted invariably hinges around economic considerations.

Sensible Heat Gains occur due to conduction, convection and radiation of heat from a material at a higher temperature than its surroundings.

Latent Heat Gains occur due to evaporation or moisture gain from a surface at a higher temperature and vapour pressure than its surroundings.

Ventilation Rate, or infiltration, controlled or otherwise, produces both sensible and latent heat gains.

An appreciation of the implications of the above three factors in discussion with the design engineer should, however, enable the architect to establish the importance of the degree of conditioning that is necessary for a given type of occupation and, in consequence, to establish the relative importance of costs that must be allocated to the various aspects of the building, as a whole, if it is to function satisfactorily when completed. Taking this reasoning to extremes: if full conditioning is *essential* to a project and an overall cost has been allocated, the client may, to meet these conditions, have to face the fact that he cannot have the standard or quantity of accommodation which he has considered necessary, assuming that the architect has been unable to do this within the limits of the remaining costs. The alternatives must be to lower the standard of accommodation, develop in stages, or reduce the degree of conditioning. All too often the cost of the fabric, and the 'degree' of conditioning are incompatible; but this is not established until an advanced stage of planning has been reached. This results in having subsequently to re-appraise the relative importance of the various aspects of the building, or to abandon the scheme. To enable the architect to establish orientational or zonal effects on his overall scheme at the outset, it is possible, as explained below, to calculate solar gains.

Calculation of solar gains

In order to reduce the effect of solar gains on the building, it is worth carrying out 'orientation-shape' exercises to ascertain if there is any significance in the shape of a proposed building when related to a particular site; and, if so, what the implications of changing the shape or reducing the solar effects may be.

To do this it is necessary to use a sun-chart or solar-progression chart. On this chart, an outline building shape is superimposed and the implications of its orientation can be seen. The solar-progression chart (Chart 1, *Appendix* A) is an example applicable to a typical location in the U.K. There are also other charts currently in use prepared by the Building Research Station and the I.H.V.E. The various angles referred to by design engineers are indicated in Fig. 14.1, and, from information derived from Table 6.3 in the *I.H.V.E. Guide*, the sun position at any time of day, and at latitudes 0° to 55°, can be plotted and

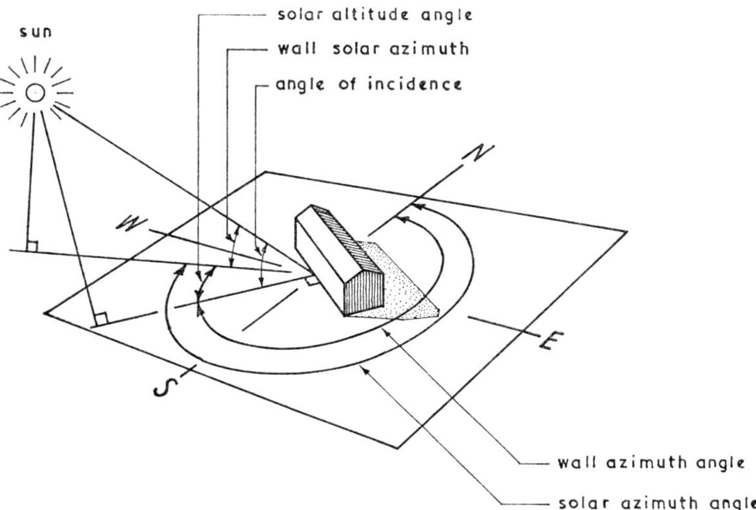

FIG. 14.1.   Diagram of Solar Angles.

its effect on building shape assessed. This gives an indication of
the overall magnitude of the solar effect on a particular building
shape in a particular location and establishes the thermal
zones resulting from this. To arrange how to offset this effect,
if the shape as a whole cannot be reorientated, needs consideration of the element or typical elements forming the building
complex. By consideration of the element, it is possible to see
the shade effects produced, and the summation of these shaded
areas would be applicable to the building as a whole. In order to
do this, Chart 5, *Appendix* A, shows a graphical method of
assessment. Variables of 'L' (depth of window recess), 'W'
(window width), and 'H' (height of windows) can be applied.
But there will be limitations to the effectiveness of such treatment, and thought must then be given to the secondary means of
offsetting direct solar gain by the use of blinds, curtains or
special non-actinic glass. For further detailed study, a series of
sun-path diagrams have been produced in *Research Series 39*,
by P. Petherbridge, M.Sc., F.I.E.S., for the Building Research
Station.

15: The Fully-Controlled Environment

1. Basic considerations

THERE IS AN ELEMENTARY APPROACH which, on consideration of a few basic requirements, will broadly indicate the types of system that are worth further consideration (possibly in some detail) by the student.

1. Firstly, the 'degree' of comfort conditioning that is *essential* to the proper functioning of the occupants within a space – e.g. (*a*) the operating theatre or (*b*) the office block.

The conclusion must be drawn immediately that space-for-space the cost of providing for the former, and its plant allocation, must be higher than for the latter, and that, of necessity, an all-air system is more than likely to be required than a combination system of part air and part water as would be suitable for the office block.

2. Whether, or not, re-circulation of the treated air is acceptable within the occupied space or element or the building as a whole.

In the case of 'Barrier Nursing' – or its equivalent of highly contagious or highly clean areas – the incoming air must always be 100 per cent outside air, pre-treated and conditioned, whilst the discharged air must not be allowed to re-circulate by accident or design. On the other hand, the requirement for a theatre or restaurant, or a place of high-density occupation, is a minimum statutory ventilation fresh-air requirement – the bulk of the air within the space being re-circulated, mixed with fresh air, and re-treated for re-circulation. Also, there is the closely-controlled working space within which few people work, with a minimum ventilation fresh-air requirement and a maximum re-circulation of room air. Finally, there is the office space in which an optimum supply of ventilation fresh air is required. Maximum re-circulation of room air is acceptable within the space; minimum use of fresh air permits this to be lost by exfiltration, without the necessity for re-circulation back to the central plant.

The types of system which would be acceptable range through 100 per cent air supply and extract: the all-air system with maximum fresh air and minimum re-circulation; the minimum fresh-air requirement and maximum re-circulation requiring high-cooling load; and finally, for the maximum space re-circulation and minimum fresh-air requirement with high sensible cooling, the combination system is applicable, rather than the all-air system.

Broadly speaking, therefore, the smaller the ratios of

$$(1) \ \frac{\text{Total Air Circulated}}{\text{Fresh Air Required}} \quad \text{and} \quad (2) \ \frac{\text{Sensible Heat Gains}}{\text{Latent Heat Gains}}$$

to meet specific requirements, the more is the likelihood that it can only be met with an all-air system – which, in turn, implies that a greater facility for service ducts will be required. Conversely, the opposite can also be said: that the greater the ratio of (1) and (2), then the more is the likelihood that a system of chilled water with localized fresh-air inlet to the units would be acceptable.

Between these two systems lie a number of combinations of water/air systems which can be considered to meet the varying demands of aesthetics, both internally and externally, of economics and also of practicability for whatever the particular type of building may happen to be.

2. Secondary heat transfer

Listed below are some of the available means of cooling building spaces:

A. *Systems providing the individual means of cooling and heating within elements*

Type of System	Associated Plant	Application
(i) Room conditioners (air-cooled condensers)	—Self contained	Domestic and private offices
(ii) Packaged air conditioners (local or remote condensers)	—Self contained	Commercial, industrial, laboratories

B. *Systems capable of providing individual heating or cooling within zones*
(with primary distribution of heating or cooling water from a central plant)

Type of System	Associated Plant	Application
(iii) Room conditioners (water-cooled condensers)	—	Commercial offices, Large residential accommodation
(iv) Fan-coil units	—	ditto
(v) Packaged air-handling units	—	Commercial, Industrial

All these systems can be provided with direct localized fresh-air inlets for ventilation purposes in the simplest form.

C. *Single and dual systems served from a central plant* (using all-air or combination of primary air and heating or cooling water)

Type of System	Associated Plant	Application
(vi) Fan-coil units	—(a) Primary ducted air for ventilation requirements only. (b) Primary media of piped hot and cold water to unit.	Commercial, Industrial, etc.
(vii) Induction units	—(a) Primary high-velocity air ducted to each unit. (b) Primary piped hot or cold water to unit.	ditto
(viii) Suspended metal pan ceiling (acoustic/ heated/ cooled)	—(a) Primary ducted air either low- or high-velocity systems. (b) Primary piped hot or cold water to ceiling grid.	ditto
(ix) Traditional low-velocity single-duct air supply to outlet diffusers, etc.	—Central plant providing heated/cooled/air.	General application (re-heating and cooling coils required for element or zonal requirements)
(x) Packaged air-handling units	—(a) Primary low-velocity ducted air for ventilation to each unit. (b) Primary piped hot and cold water to units.	Large multi-zone installation

Type of System	*Associated Plant*	*Application*
(xi) High-velocity single-duct supply to attenuator boxes and outlet diffusers	—Central plant providing heated/cooled air, requires re-heat or cooler batteries for zonal control.	Commercial, Industrial
(xii) Twin-duct high-velocity ducted air to mixing boxes and outlet	—Central plant providing heated and chilled air before discharge at desired condition into space.	General application

It is fairly obvious, from even the above brief summary of the different systems, that the final design and recommendation for the most suitable type of system will be outside the scope of the architect's training. However, with his experience and technical appreciation, the designer will be able to advise the architect on the most economical and efficient installation to meet the specific problem.

The extent to which the architect appreciates the degree of internal conditions required, the purpose for which the building is to be used, the effects of orientation within the limits of the site, and the anticipated methods and materials to be used by him in the construction, will influence to a great extent which system can be used and will reflect in the initial cost and the running costs. With such a responsibility, it is naïve to suggest that any non-specialist designer, however well trained, would be capable of assessing the parameters which should be applied in a complex situation.

Apart from the associated structural and aesthetic factors, these facts alone should serve to show that it can never be too early in the planning of a new building of reasonable complexity for the architect to bring together all the knowledge and experience of others, before committing his interpretation of these factors to paper.

In order to assimilate the information which will be put to him as an architect, the student should endeavour to appreciate some of the fundamentals which will influence the design engineer's approach to the problem.

System Selection

Firstly, if the building can be so constructed and orientated that the external effects of climate are minimal, this is the first objective to be aimed at, as outlined in the foregoing chapters.

In theory, if the structure were such that its thermal inertia prevented heat gains passing-through during normal occupation, and, in addition, if the glazing was shaded so that no direct solar gains occurred, there would be little occasion for cooling in temperate climates. However, let us consider the normal conditions in which there are heat gains. If the requirements for 'conditioning' are local, and not too exacting, then either of systems (i) or (ii) could be adopted – as, for example, the individual private office, the private home, the board-room or the small store, depending upon the size.

Increasing the scale of requirements to include 'conditioned' multi-storey flats, small offices, mechanized offices and the like, systems (iii) to (v) would be applicable.

The next stage might be considered where a 'degree' of conditioning is *essential*, but must be 'low cost'. For example, for barracks overseas or for commercial developments, then systems (v), (vi) and (vii) could be considered. These would begin to have limitations inasmuch as humidity could not be controlled, somewhat high noise levels would have to be accepted, and the possibility of under-cooling or over-heating might occur at the seasonal change-over.

Up to this stage, the types of system used would, in the main, be capable of providing the cooling requirements to deal with sensible gains, relatively low ventilation rates and a degree of dehumidification; beyond this stage, the necessity of systems capable of handling humidification, high ventilation rates and large sensible cooling loads, with low noise levels, high filtration of dust and bacteria from the air, demands more refined installations.

Finally, therefore, hospitals, prestige office buildings, large machine rooms, theatres, hotels, etc., would be considered as needing either one of, or a combination of, systems (vii) to (xii).

The detailed methods of distribution are outside the scope of this book, and the student would be advised to refer to the more comprehensive information which is available for the particular installations, (i), (ii) and (iii) (see the *Bibliography*, page 216).

Reducing Heat Gains

It should be clear that in all these systems the uncontrollable internal-heat gains derive from plant, or machines, from personnel and lighting, and they have to be accepted as providing a basic level of requirement for cooling. Any solar-heat gains resulting from orientation and structural design which can be eliminated, or reduced, should be dealt with by design or selection in the early stages of planning the building.

Lighting can represent a significant heat-gains problem in certain circumstances: for example, in a large department store it can represent approximately 33 per cent of the total heat gains. The full effect of heat gains from lighting within a space can be reduced with recessed-type fittings and the combined light-fitting/air diffuser. This lays stress on the fundamental approach that the *relative magnitude* of the different services and elements within the building should be considered at an early stage, in order to assess the 'priorities' for a particular building and its design function.

For example, the technique of lighting, if not handled with some expertise, can increase the cooling-load requirements in achieving a stated level of illumination. Yet the same 'degree' of illumination could, when handled by a competent designer in collaboration with the architect, be provided with less actual heat gain into the space. Similarly, the provision of natural or day-lighting is not always incompatible with the thermal requirements, since both attempt to reduce direct sun-effect.

Competent design, integrating natural light and artificial lighting, results in an overall saving on the internal heat gains.

We have already considered the effects of solar-heat gains, and have seen that they are to a certain extent controllable. For example, to place an insulating-quilt in the roof construction is a palliative, but, as explained in an early chapter, a reflective membrane will throw-back the radiant heat more effectively than the quilt forms of insulation. In consequence, a combination of these two would produce the right effect. Similarly, while roller blinds and venetian blinds do have an effect on solar gain, they also tend to become hot themselves and begin to re-radiate the heat into the building. For example, in regions in which the climate produces the problem of constant rather than an occasional solar gain, there is more logic in the use of projecting sun-breakers, formed as part of the structure and carefully designed to present shaded surfaces toward the building, yet retaining a form which can dissipate absorbed heat by conduction and convection to the surrounding external air. Wide-spreading eaves or overhanging balconies designed to

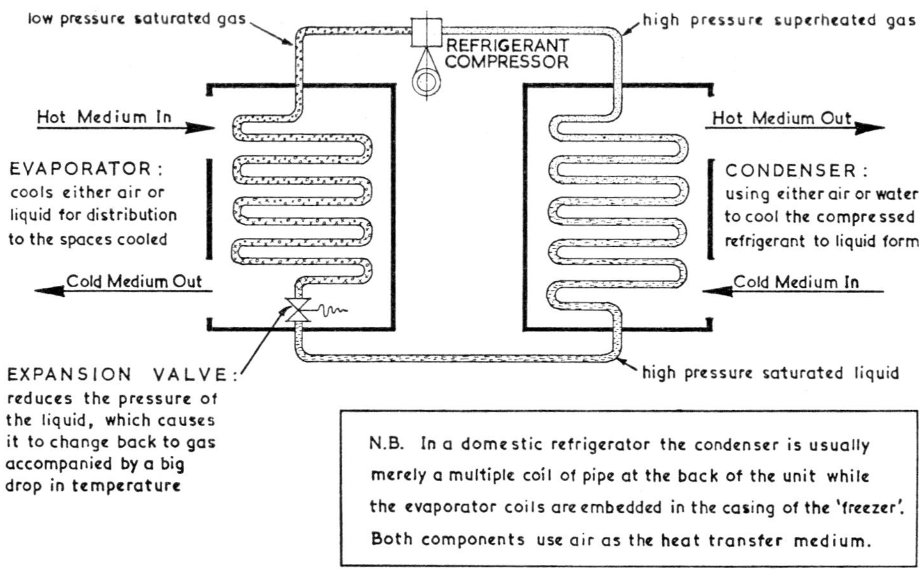

FIG. 15.1.   Simple Mechanical Refrigeration System.

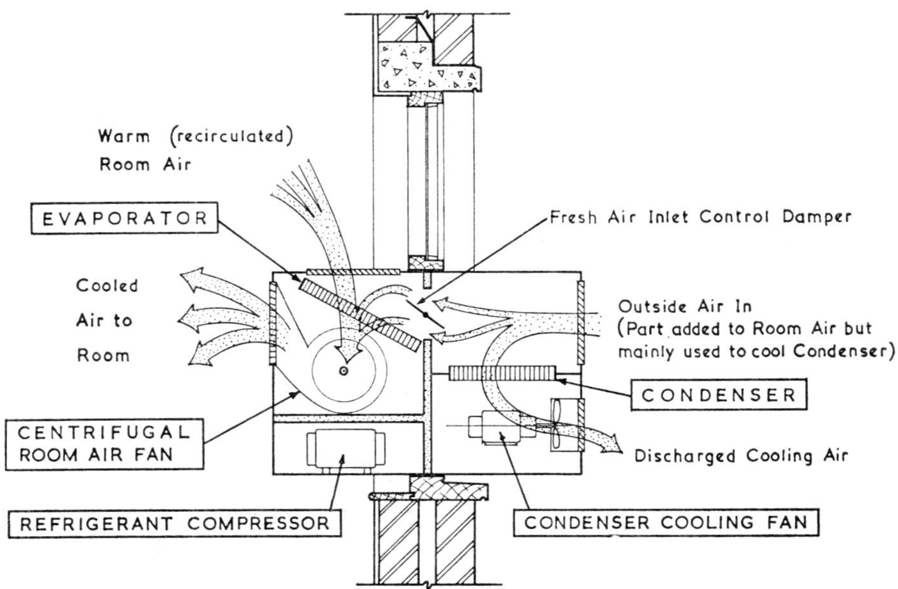

FIG. 15.2.   Extension of Basic Systems for a Local Room Conditioner.

shade windows, as in 'ranch-type' verandah dwellings, and white-painted reflective surfaces are examples of the influence of the thermal requirements on the architecture and structure of the buildings.

3. Primary generation—cooling

Later on in this chapter are indicated the plant areas that need to be considered when buildings require more complex services. Estimates of plant area were given in Chapter 11 for the primary plant.

Let us consider the essential elements of an air-conditioning system, bearing in mind that heating-plant space is still required to provide the primary-heat generation as well as the cold-generation plant.

In its simplest form, the cold generator operates in exactly the same way as a domestic refrigerator, except that the unit cools the air or other media which, in turn, is distributed and absorbs heat from the conditioned space. The basic components are therefore simple and, in grasping these, an appreciation of the larger components of the central plant can be more easily obtained.

Elementary Principles: Fig. 15.1 shows the basic components for a simple mechanical refrigeration cycle (cold generator reducing a fluid to a low temperature).

To adapt this for a 'room-conditioner', the condensing medium is the outside air, while the evaporator cools the room air which is drawn over the evaporator by means of a fan (Fig. 15.2).

The larger unit, which cannot be mounted in a window, can consist of a floor-mounted evaporator and compressor unit, with a separate condenser unit mounted externally, as shown in Fig. 15.3. Alternatively, to obviate the connecting refrigerant lines, all the elements are concentrated in the one floor-mounted unit while condenser cooling air from outside is ducted to and from the condenser.

For centralized systems, much the same components are required, but are sited in plant rooms which are distributed about the building.

Broadly speaking, there are four basic components to provide complete air-conditioning systems from (v) to (xii). These are as follows:

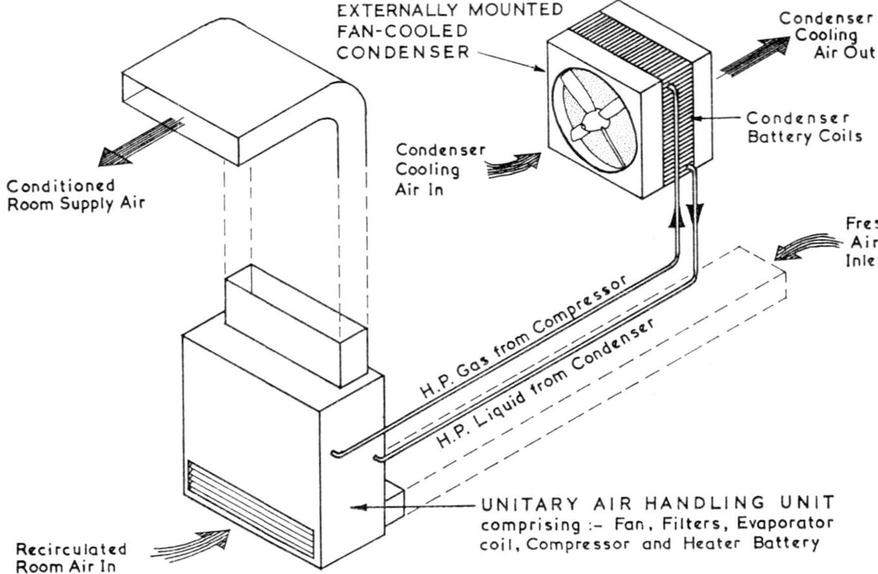

Fig. 15.3.   Floor-Mounted Conditioner.

1. *Primary Heat-Generation Plant* – described in previous chapters.

2. *The Refrigerant-Cooling Plant*, used either to provide direct-expansion cooling or indirect-cooling coils *via* chilled water, comprising refrigerant compressor, condenser and evaporator.

3. *Air-Handling Unit*, in which the air is 'conditioned' and includes air filters, heater coils, cooling coils, air washer or sprayed coil for humidification control. These can be in individual units, purpose-designed and assembled, or unit assemblies within a specific-range of standard units.

4. *Terminal Units*, used solely with item (1) and (2) or in conjunction with (3) to provide the combination or dual system.

Let us consider the individual components outlined above and illustrated in Fig. 15.5 (see page 186).

4. Refrigerant cooling plant

Refrigerant Compressor

Refrigerant compressors can be mechanical centrifugal machines or reciprocating machines, chemical absorption machines, or thermal steam-jet. The detailed description of the different types would require further investigation into reference books or trade literature.

The factors which are particularly relevant to the architect are:

(*a*) noise

(*b*) location

(*c*) accessibility

(*d*) size

(*e*) interrelation of the various components.

The electrical and mechanical compressors are relatively heavy and noisy and, in consequence, are normally mounted in the basement; they require ventilation, and care needs to be taken to avoid disturbance from the resultant noise. Chemical absorption machines are quiet, relatively large and require to be located near to a primary source of heat, such as a steam or high-temperature hot-water boiler. Steam-jet refrigeration can only be applied when suitable steam-boiler plant is also an integral part of the services.

Accessibility is required initially to get the plant in, or subsequently out, and for adequate routine maintenance.

Condenser

This can be an integral part of the refrigeration plant, as in a shell and tube water-cooled condenser, or remote from the compressor, as are air-cooled condensers. If the condensers are of the shell and tube type, water-cooling towers are required, unless an adequate water supply is available as in the case of an adjoining river. Water-cooling towers or condensers are normally mounted on the roof and can produce problems due to noise and vapour and, in consequence, their location must be considered in relation to adjoining buildings.

Evaporative Chillers

These can be an integral part of the refrigeration plant and provide the means by which the media of air or water are cooled. They can either directly cool the air, as in a direct-expansion cooling coil in the air-handling plant, or in a bulk-storage chilled-water tank as an integral part of the air washer, or they can produce chilled water distributed to cooling coils in the air-handling plant.

Pipework

In the case of air-cooled condensers located on the roof, compressors in the basement and remote air-handling plant for each zone, the distribution pipework between these items is usually large in size due to lagging, etc. The most direct route between these items is required for the services ducts which contain the pipework.

5. Air-handling unit

Current practice to speed-up building operations has put a greater emphasis on factory-assembled equipment which will reduce on-site work. In consequence, unitary factory-assembled components of the basic air-handling plant – namely, fan, motor, filter, heating and cooling coils and humidifiers – are available to cover a wide range of applications, ranging from 0·5 m³/sec to 19 m³/sec (1000 cfm to 40,000 cfm). This unit, or

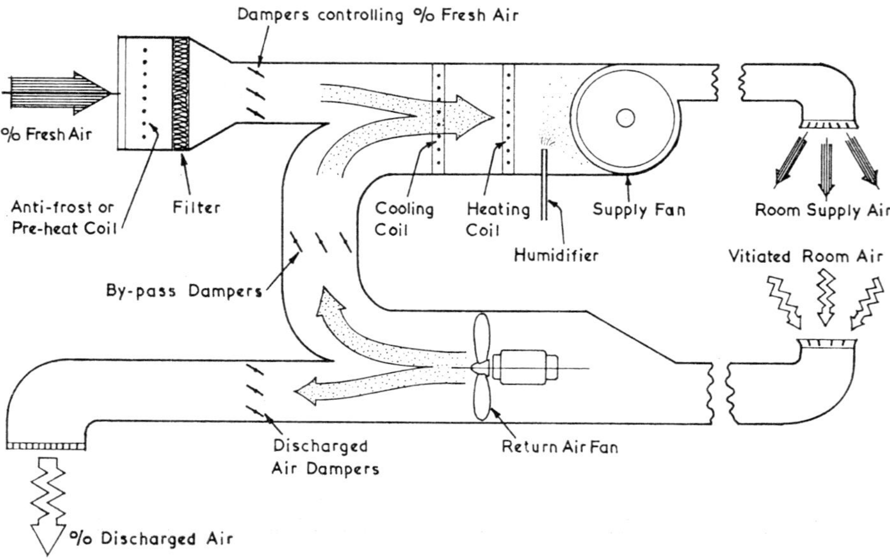

FIG. 15.4.   Diagrammatic Layout of Air-Handling Plant.

its separate components, site-assembled, provides for primary-air supply. Depending upon the extent of the recirculation of conditioned air – or even when the air cannot be recirculated (as in the case of a hospital), but is completely discharged – a fan or fans are required in addition to the supply air-handling equipment. This is diagrammatically shown in Fig. 15.4.

Unitary systems are available up to 19 m³/sec (40,000 cfm), but, thereafter, they begin to become an integral part of a Central Station. For example, fresh-air inlets for a large installation are large too. A typical calculation might be as follows:

$$\frac{19 \text{ m}^3/\text{sec}}{2 \cdot 5 \text{ m/sec}} = 7 \cdot 6 \text{ m}^2 \left(\frac{40,000 \text{ ft}^3/\text{min}}{500 \text{ ft/min}} = 80 \text{ sq. ft} \right),$$

i.e. the opening is 3 m × 2·6 m

Such elements for this size of plant begin to need integrating as structural elements of the building. The figure on page 186 (Fig. 15.5) showing the size range of these various components of the completely conditioned building, will serve to indicate the allowances which may have to be made to accommodate the items of plant.

Space allocation is indicated pictorially for a nominal range of air-handling plant from 0·5 m³/sec (1000 cfm) to 22·0 m³/sec (48,000 cfm); refrigeration plant, allowing for compressor/chiller and air-cooled condenser, is within the range of 45 kW to 380 kW (12 to 100 tons of refrigeration). (The obsolete unit of a ton refrigeration represents 12,000 Btu/hr.)

It must be borne in mind that each of the units shown in Fig. 15.5 requires 'working space' for maintenance and plant replacement; in consequence, access openings equivalent to a size of a unit plus clearance should be allowed for. Similarly, adequate headroom is required above the units, particularly the air-handling unit, to facilitate the installation and distribution of ducting, pipework, etc. In addition to the basic units, there are ancillary items of equipment such as pumps and control panels, quite apart from the space required for the primary-heat generation plant, which may be the boiler plant complete, or simply a heat exchanger served from remote central heat-generating plant.

There are also numerous combinations of plant and varying types of plant which can be used to achieve an equivalent result,

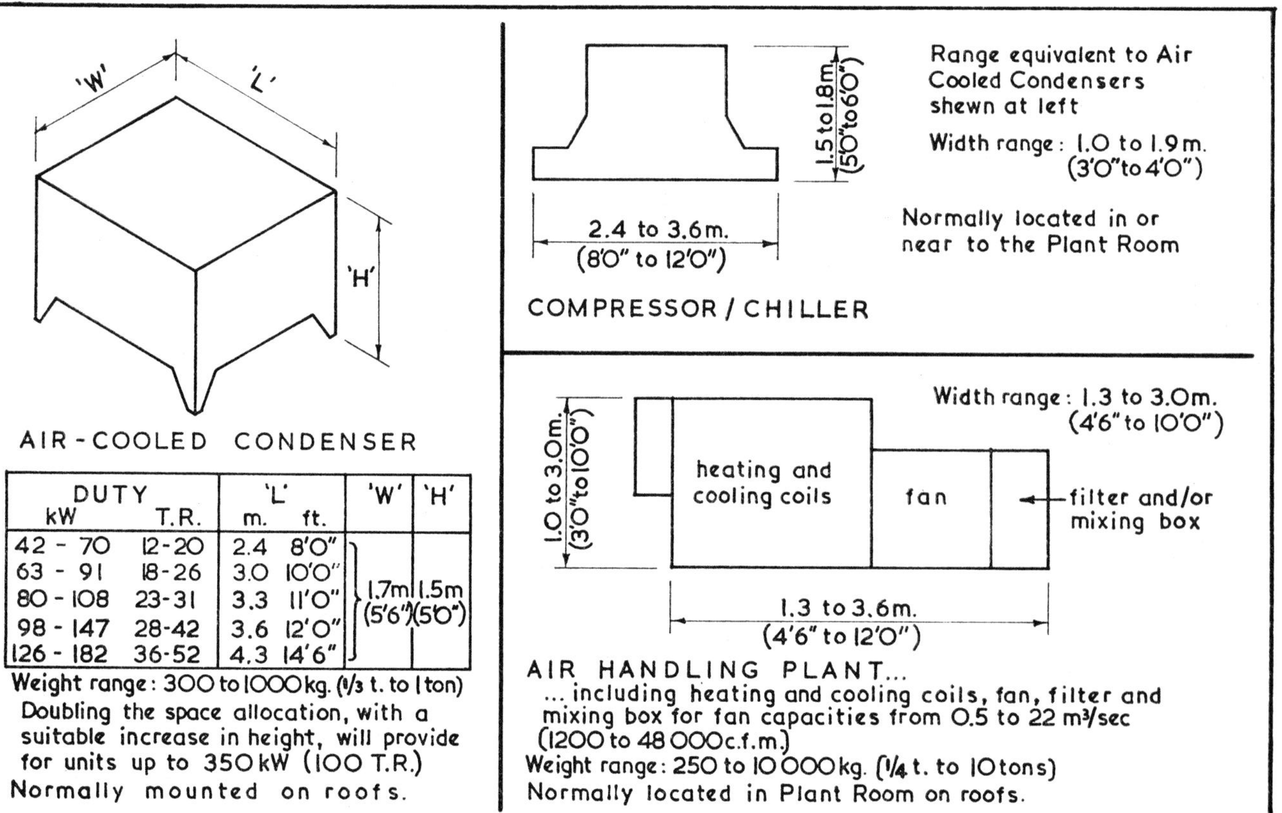

| DUTY | | 'L' | | 'W' | 'H' |
kW	T.R.	m.	ft.		
42 – 70	12-20	2.4	8'0"		
63 – 91	18-26	3.0	10'0"		
80 – 108	23-31	3.3	11'0"	1.7m	1.5m
98 – 147	28-42	3.6	12'0"	(5'6")	(5'0")
126 – 182	36-52	4.3	14'6"		

Fig. 15.5.   Space Allocation for Air-Handling Plant.

and it is obvious, therefore, that at the early stages of space allocation, when only broad principles are being agreed, the allocation of the appropriate 'boxes' within the building complex is the best that can be achieved. This is probably particularly so in the student's case where time or facilities do not permit him to investigate the problem fully, or where he does not have the facility of obtaining information or advice from the designer of the services.

Bearing this in mind, the following synopsis of typical space requirements has been prepared and, whilst it is an over-simplified approach to the actual problem, it is hoped that it will provide the student with the facility of allocating some spaces in juxtaposition to each other and to the building complex as a whole. Also, as has been mentioned previously, the overall space allocation is governed by the amount of plant that has to be accommodated, and this in turn is influenced by any requirement for 'stand-by' plant for a given functional use of a building.

Allocation of Plant Space for Air-Handling Equipment and Refrigeration Plant

Building Cube ..	1400 to 7000 m³ (50,000 to 250,000 ft³)	8400 to 28,000 m³ (300,000 to 1,000,000 ft³)
Air-Handling Plant (equivalent floor area expressed as % of building cube) ..	0·18% to 0·25%	0·15% to 0·2%
Refrigeration Plant (expressed as a % of the air-handling floor space)	75% to	25%
Electrical Switchgear (expressed as a % of the air-handling floor space)	20% to	33⅓%

There is a minimum height of approximately 3·0 m (10 ft 0 in) for plant rooms requiring ducting distribution, and this extends upwards to 8·0 m (28 ft 0 in) for the larger or more complex plant rooms.

The refrigeration plant tends to require proportionately less space, the larger the installation; in consequence, a greater percentage of floor space should be allocated to the smaller building.

On air-handling plant space, it can be assumed at this stage of planning that the choice of a low-velocity conventional system or a dual-duct high-velocity system will require a large percentage of floor area, as, in all probability, the one plant will provide for all the heating and cooling required. Whereas a single-duct, high-velocity system will probably be used in conjunction with the terminal units itemised as (vi), (vii) and (viii) at the beginning of this chapter.

From the foregoing, it should be possible to estimate the 'boxes' required for a given building cube and its associated thermal characteristics. For example, a building of given cube – with a primary fuel of oil being most appropriate, in a location requiring sealed windows, a relatively high machinery load, artificial lighting and high density of occupation – will require space allocation for services as indicated in Chapter 13 for oil-fired primary-heat generation, and as indicated in this chapter for primary-cooling plant (air-handling plant ventilation, cooling and associated refrigeration space). The type of system selected as applicable might be (viii), providing a combination installation. This is a brief indication of the procedure which it is hoped the architectural student would apply in more detail, from an earlier appreciation of the 'thermal characteristics' of his building, to enable him rationally to assess the degree and implications of the services requirements. The next stage would be the more detailed consideration of 'element' and 'zonal requirements', leading finally into the sketch scheme and final-design stage.

In this chapter, consideration has been given to the primary source of generation of cooling media and the secondary means of transfer in terms of air-handling plant or terminal units. The co-ordination, feed-back of information, co-relation and adjustment of the primary and secondary sources of heat transfer are achieved through the thermostatic-control system.

6. Thermostatic control of air-conditioning systems

The control systems for air-conditioning installations are basically as described in previous chapters for heating and ventilation
and can be mono-system or a combination of all three to achieve
the control required. The control systems are obviously more
complex in order to provide for control of heating, cooling,
humidification or dehumidification, but they are mainly associated with the control within the plant room, with the exception
of the external 'pilots' and solar 'indicators' mounted externally
to 'sense' the external conditions affecting a particular zone.

Thermostatic Controls can be simply described as the *link*
between secondary heat-transfer units and primary-heating and
cooling generation plant.

The objects of controlling the system are two-fold:

(*a*) to maintain the overall thermal efficiency of the primary
plant;

(*b*) to sense the change required from the secondary heat-
transfer system in relation to external climate etc., and to
maintain comfort conditions within the occupied space.

In order to control the various 'zones', it is necessary to
locate sensing elements on the particular face of the building
relating to the zone, as mentioned above. Due to the relatively
small size of the sensing elements, this does not normally prove
any embarrassment to the architectural treatment of the
building.

7. Plant location and accessibility

Broadly speaking, the elements of constructional and architectural design are 'static', whereas the elements of the engineering services are 'dynamic' and, in consequence, are subject to
periodic adjustment and deterioration with time and to subsequent obsolescence and replacement.

Generally, therefore, as systems become more complex so
they require greater accessibility for mechanical and electrical
maintenance and adjustment. Whilst it is not possible to
decide the full extent of such requirements until a fairly advanced stage of design, awareness of the problem and consideration of this aspect in the early stages of planning must ultimately reflect in the ease with which the various stages of the work
progress and in the final successful functioning of the building.

(*Continued on page 192*)

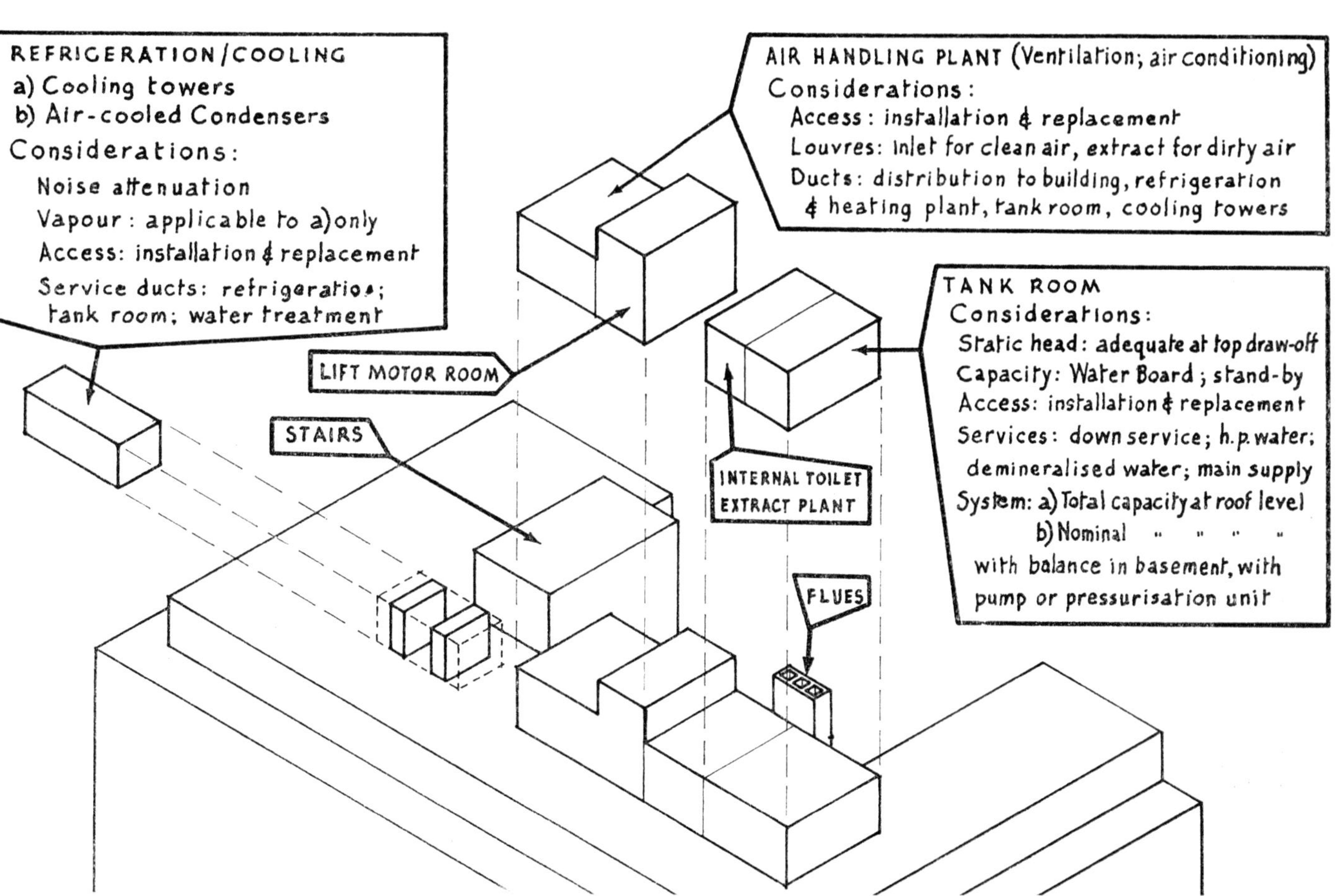
REFRIGERATION/COOLING
a) Cooling towers
b) Air-cooled Condensers
Considerations:
Noise attenuation
Vapour: applicable to a) only
Access: installation & replacement
Service ducts: refrigeration; tank room; water treatment
AIR HANDLING PLANT (Ventilation; air conditioning)
Considerations:
Access: installation & replacement
Louvres: inlet for clean air, extract for dirty air
Ducts: distribution to building, refrigeration & heating plant, tank room, cooling towers
TANK ROOM
Considerations:
Static head: adequate at top draw-off
Capacity: Water Board; stand-by
Access: installation & replacement
Services: down service; h.p. water; demineralised water; main supply
System: a) Total capacity at roof level
b) Nominal "   "   "   "
with balance in basement, with pump or pressurisation unit
LIFT MOTOR ROOM
STAIRS
INTERNAL TOILET EXTRACT PLANT
FLUES

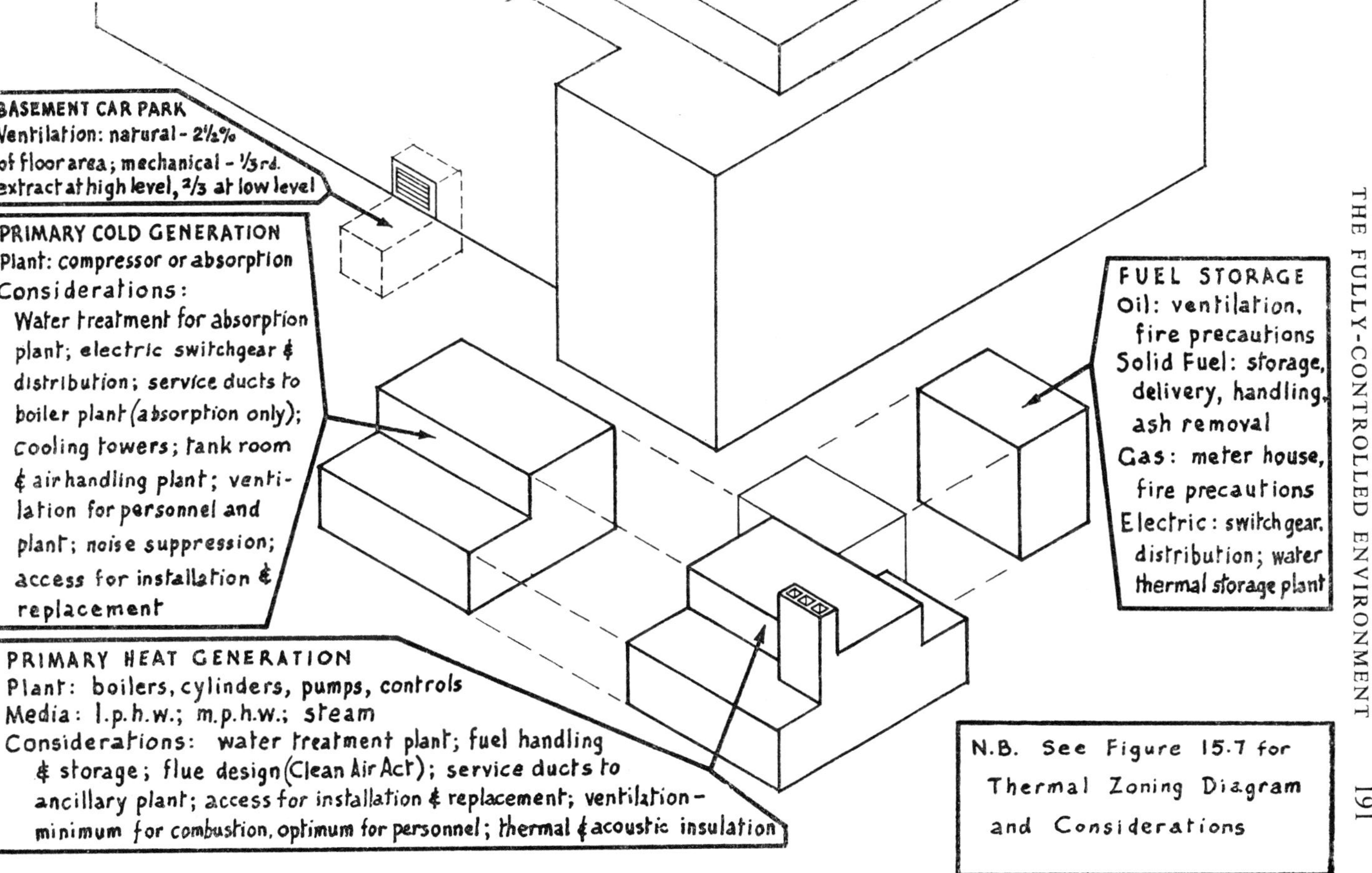

FIG. 15.6.   General Considerations concerning Plant located in a Building (see page 193).

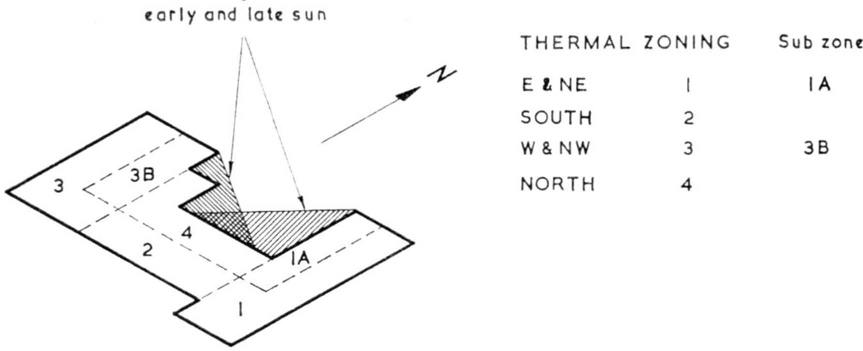

Assessment of thermal zoning will facilitate
decision on location of service ducts within
the building for the particular type of
secondary heat transfer equipment (heating,
ventilation or air conditioning).

Fig. 15.7.   Thermal Zoning of Structure in Fig. 15.6.

It is no use providing adequate service ducts and plant rooms unless the equipment can be satisfactorily installed and handled into position during construction. All mechanical plant is subject to wear and will need replacing sooner or later. The larger the installation, the larger the facilities for access must be. In particular applications such as hospitals, maintenance is vital to the functioning of the building and must be considered as an integral part of each design stage.

Provision for attentuation of plant noise is also necessary as the installations become more complex. Anti-vibration mountings are required for moving machinery, or even double floor construction in which large items of plant can 'float' on cork or rubber over the structural floor slab. In the case of attenuator boxes for high-velocity single-duct systems, or volume control boxes in the twin-duct systems, the engineer can be expected to make provision for the correct noise attenuation; but the architect is required to provide access for control and adjustment of such units within, say, the design of his false ceilings.

Fig. 15.6 illustrates a hypothetical building with various possible plant items located about the structure. The primary heat generation plant is positioned in the basement (as is common but not obligatory) with fuel storage and flue nearby. The primary cold generation plant is housed alongside it, while the condensers and air-handling plant are mounted on the roof adjacent to the main storage and expansion tanks.

The notes, which list many of the items which may be included in these plant areas, are not exhaustive, as specialised building functions may demand all manner of process plant whose position may, for reasons of noise insulation or simplified maintenance, be preferred to be in general plant rooms. A number of statutory requirements relating to certain aspects of servicing a building are also listed.

Fig. 15.7 is a reminder of the essential early step of considering the thermal zoning of the building shown in Fig. 15.6. From such considerations the most effective and economical layout of the plant will emerge.

All other aspects of the total environment, visual, aural and spatial, have to be considered together with client preference and financial limitations, and all these must be co-ordinated to produce a building complex. Thermal environment, no less than the other aspects of environment, deserves consideration, and it is the aim of this book that the architectural student should at least be capable of assimilating and discussing – even influencing – the design engineer's solution of the problem in the most satisfactory way.

8. Overall design procedure

All of the foregoing chapters can be summarised into a general design procedure, as follows:

(a) Consider the human requirements essential to satisfactory thermal comfort for a given occupational or functional use of a building.

(b) Integrate these considerations into the other aspects of aural, spatial, and visual requirements.

(c) Consider these aspects when orientating or locating the building, and adjust the details to maintain a balance.

(d) Develop and integrate, through an appreciation of the problem, the space and zonal requirements required by, or influencing, the thermal plant.

(*e*) Reconsider in detail the thermal aspects of the thermal 'elements' produced by the overall plan in conjunction with the functional, structural and aesthetic requirements.

(*f*) Cross-check that statutory and other requirements are complied with in achieving these results.

(*g*) The final result should enable the engineer to design an installation that has the merit of economy and is an integral part of the building concept.

9. Gross energy requirement

To enable the designers of the building to arrive at a reasonable early estimate of the Total Gross Energy Requirement for the satisfactory thermal environment of a building, the following sequence of ideas and symbolic formulae may be of assistance:

(i) *Environmental Balance and Demand within the Element*

On page 57, the *thermal exchange* between a body and its surroundings (i.e. the resultant heat emission) was stated as $M - W = E + R + CV + S$.

To create an *environmental balance*, the oxygen and the ventilation requirements (O and V) must be taken into account; hence the expression for environmental balance per person becomes $M - W = (E + R + CV + S) + (O + V)$.

Thus, in an Element occupied by P persons, the total *human requirements*, are $P [(E + R + CV + S \quad (O + V)]$.

The expression $(E + R + CV + S)$ finds numerical values in Table 5 in *Appendix B* and Chart 2 in *Appendix A*.

The values of O and V are, in themselves, difficult to ascertain, but will always be satisfied by the provision of the recommended ventilation rates such as are to be found in Tables 2 and 4 in *Appendix B*.

To enable the *human requirements* to be satisfied, the *physical requirements of the enclosing structure* of the Element must also be satisfied. These are Fabric Heat Loss (or Gain) and Ventilation Heat Loss (or Gain) represented below by the symbols F and Q, and they vary according to the type of structure, orientation and function. (Q, as stated above, incorporates O and V).

Therefore, the Secondary Heat Transfer Requirement (H), modified by the Thermostatic Control (C), must equal the sum of both the human requirements and the physical requirements of the Element:

$$H \times C = P(E + R + CV + S) + (F + Q)$$

(ii) *Environmental Balance and Demand within the Zone*
The total of the Element demands in a given zone of a building depends upon the Number of Elements (N) in the zone. Therefore Zonal Demands (Z) can be represented as:

$$Z_1 = N[P(E + R + CV + S) + (F + Q)]$$

(iii) *Total Nett Demand for Heating and Cooling*
The total nett thermal requirement for the complete building arises from the summation of all zonal demands:
Primary Generation of heating and cooling $= Z_1 + Z_2 + Z_3 \ldots \ldots Z_N$

(iv) *Total Gross Demand for Heating and Cooling*
To allow for secondary losses from distribution, for the thermal inertia of the structure, and for a margin on the primary generation for exceptional requirements, a percentage is added to the Total Nett Demand.

Conclusion

There is no universal answer or definite conclusion to be drawn in respect of thermal environment, any more than there is in other aspects of human sensitivity. In consequence, there is no ultimate 'right' system which will meet and satisfy the various and sometimes incompatible requirements for human comfort. The best that can be achieved is to analyse each problem, draw intelligent and rational conclusions from the analysis and then endeavour to integrate these into the other aspects of the building environment, and, by so doing, attempt to arrive at the best solution to the overall problem of providing a well-designed building.

Appendix A: Charts

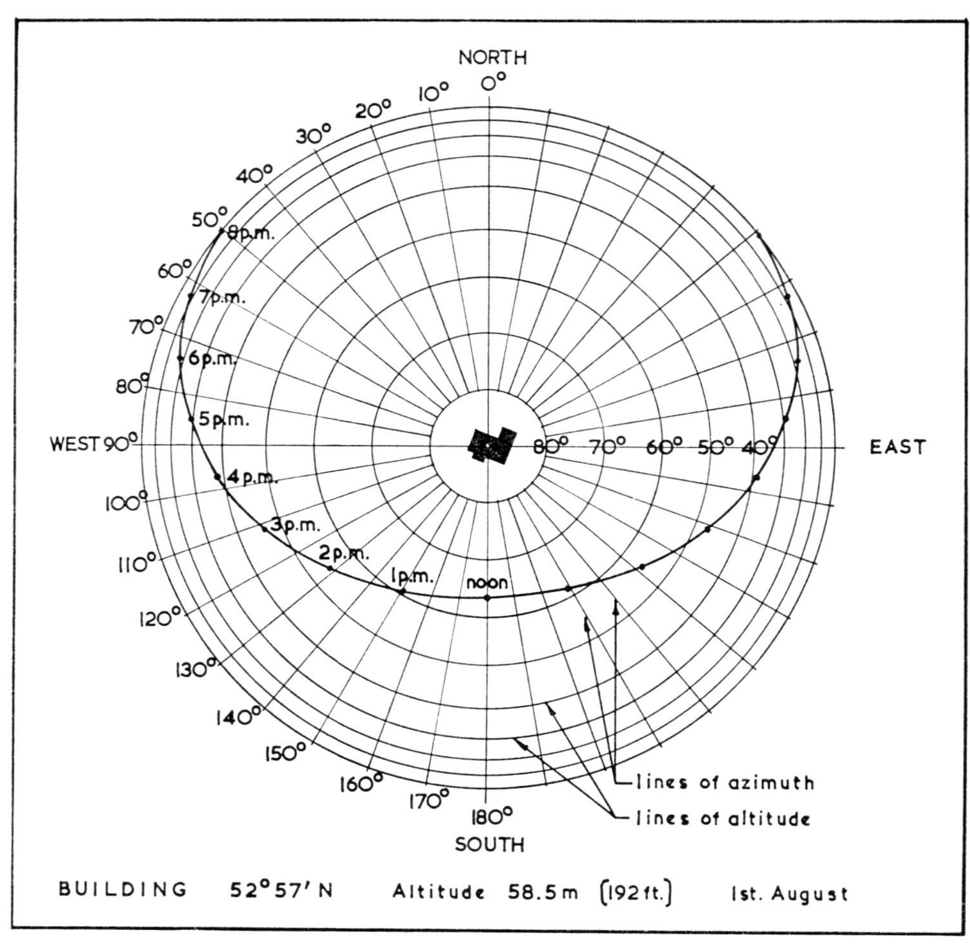

CHART 1.   Typical Sun Position Chart

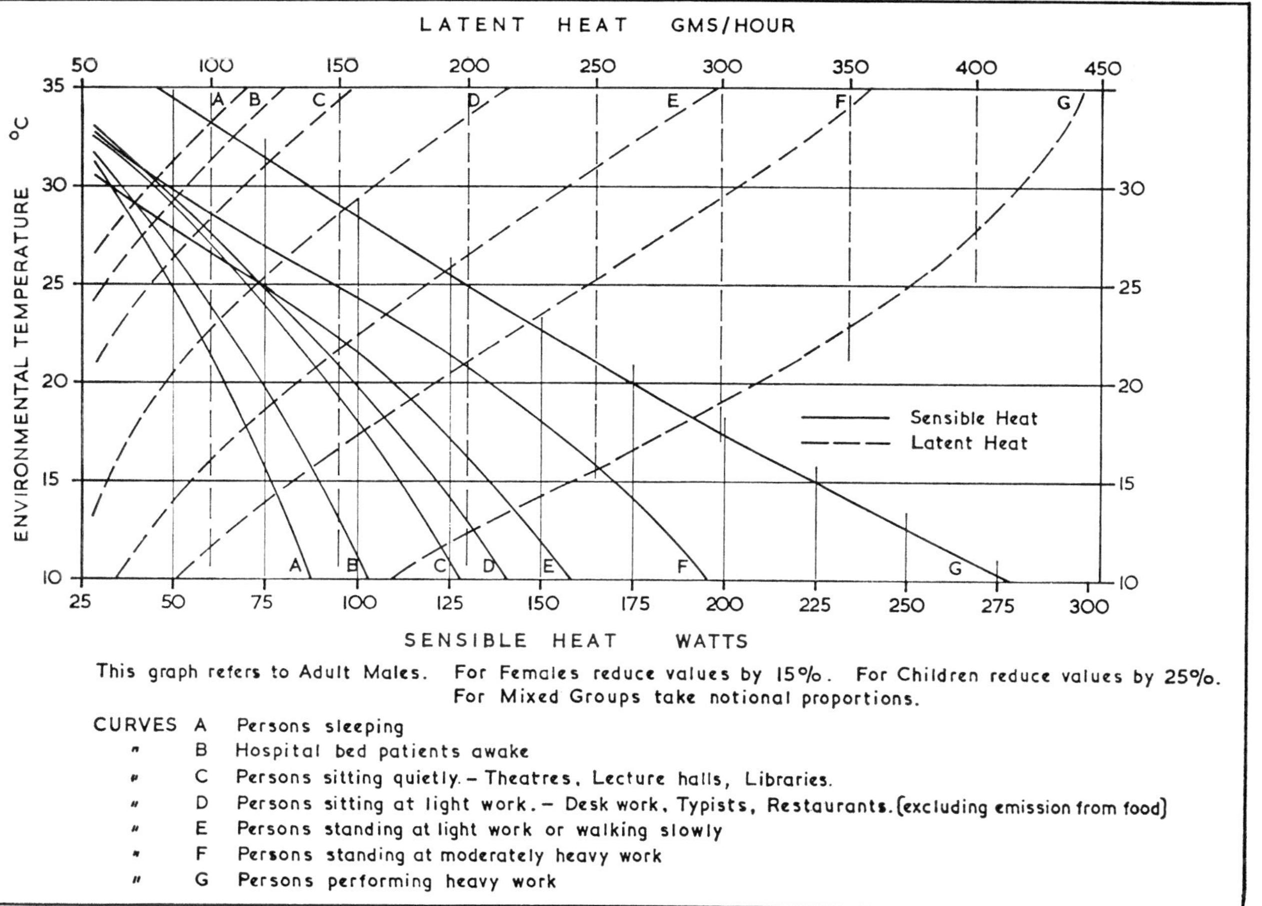

Chart 2. Heat Emission from Human Adult Males

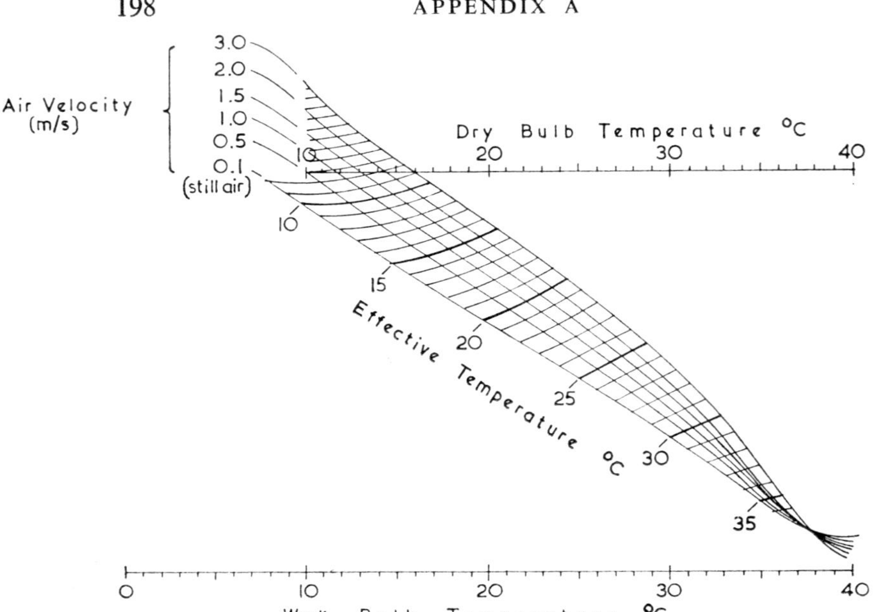

CHART 3. Effective Temperatures. Nomogram adapted from
Manuel des Industries Thermiques, published by Co. S.T.I.C.

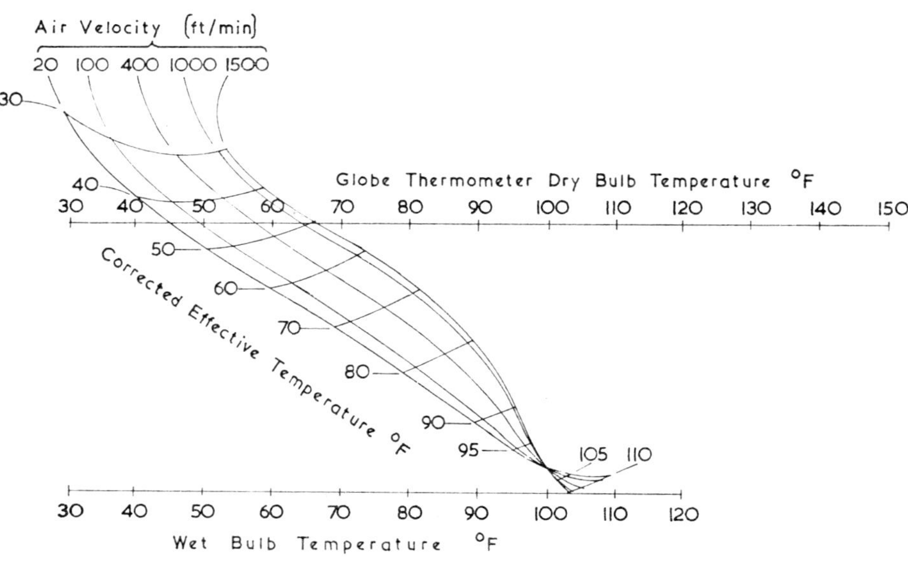

CHART 4. Normal Corrected Effective Temperature

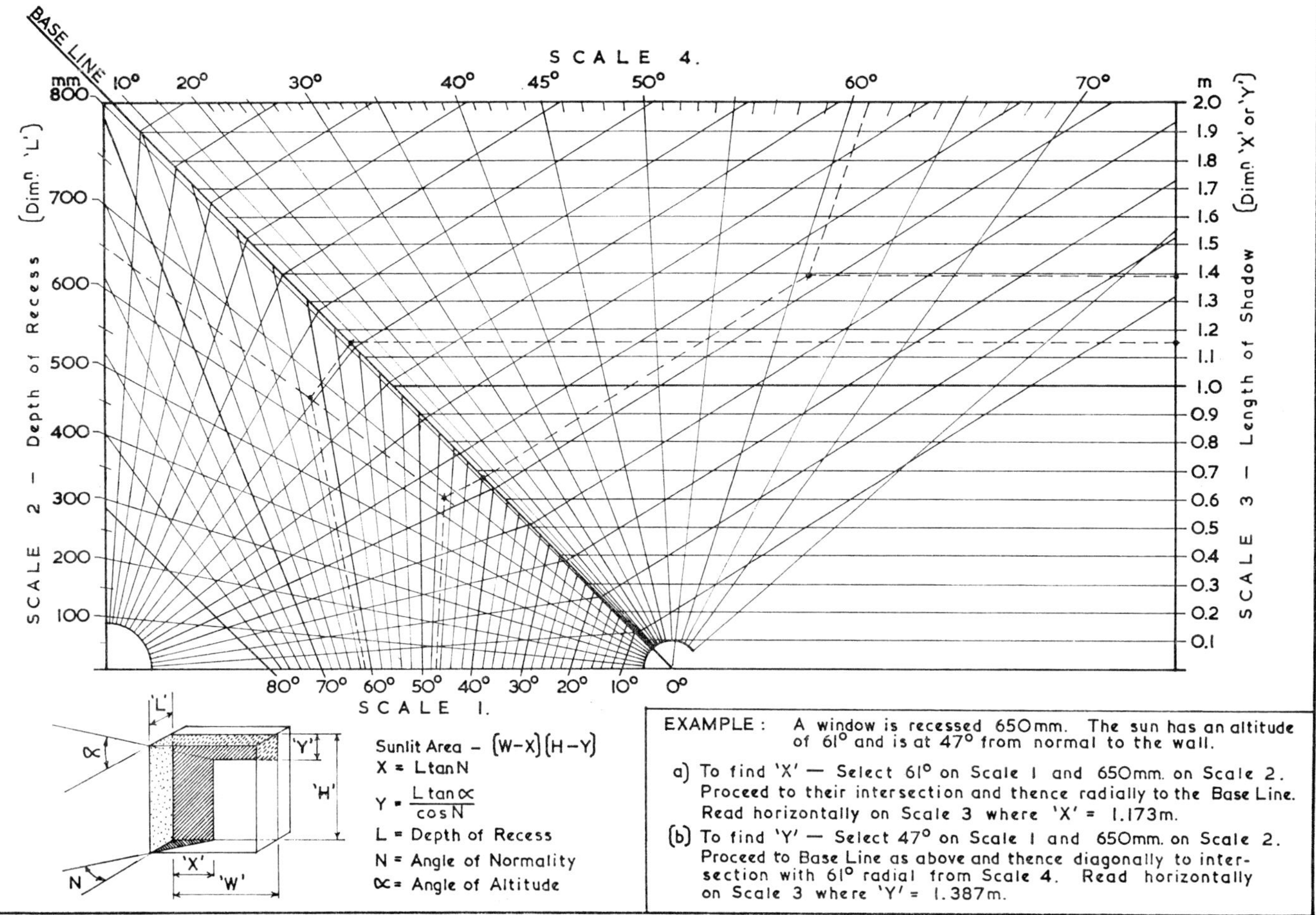

CHART 5. Window Shading Nomogram

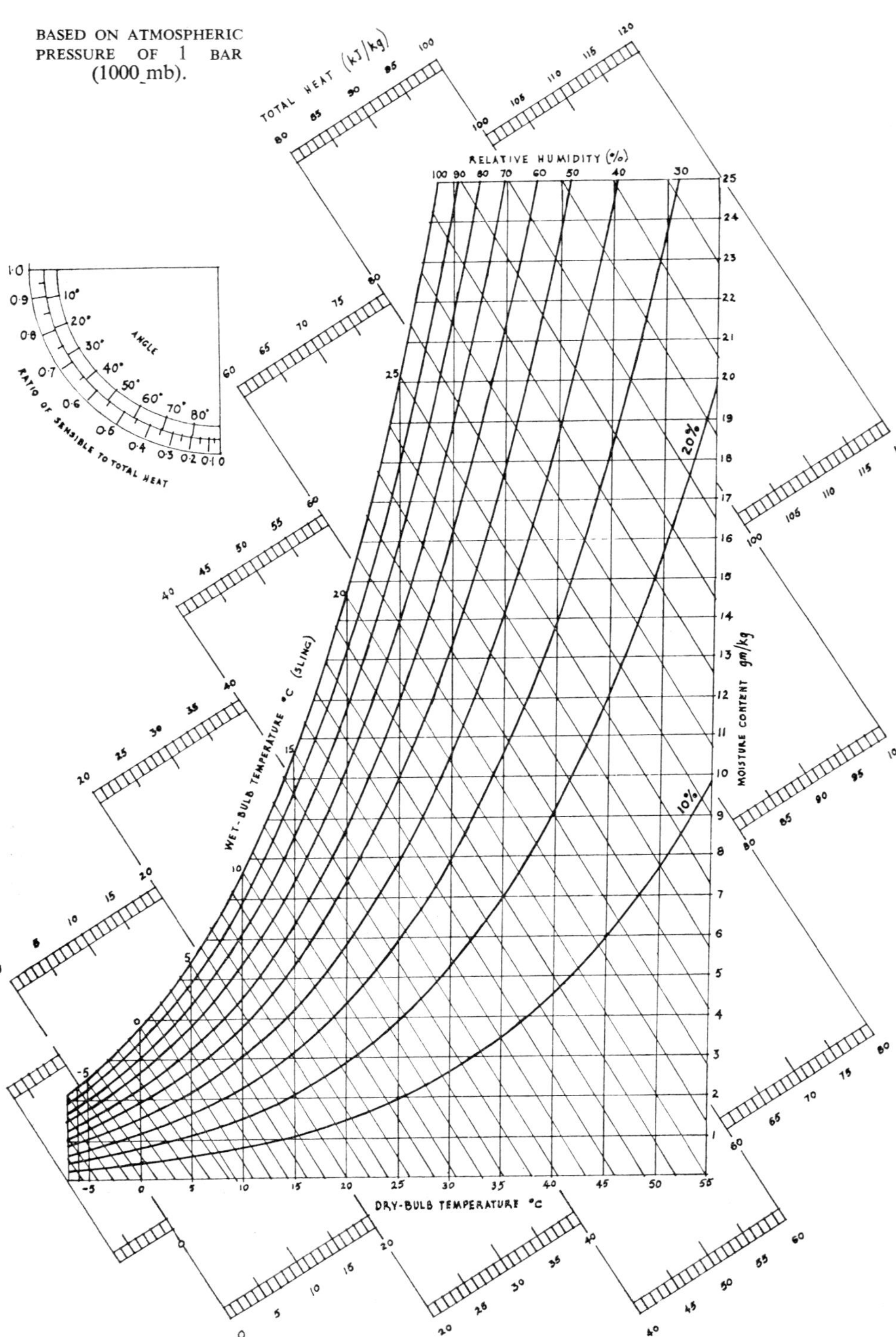

CHART 6. Psychrometric Chart (in S.I. Units)
Reference should also be made to the pyschrometric chart (in Imperial Units), published by the I.H.V.E., from which this chart has been derived.

Appendix B: Tables

TABLE I

SOME RECOMMENDED AIR TEMPERATURES

Occupancy	°C	°F
Cinemas, Theatres	15–17	60–63
Cloakrooms, Lavatories	13–14	55–57
Corridors, Staircases, Lobbies	15–17	60–63
Domestic—Living Rooms	17–19	63–67
„ —Kitchen	15–17	60–63
„ —Hall	10–13	50–55
„ —Bedrooms	15–17	60–63
Factories—Sedentary Work	18–19	65–67
„ —Light Work	15–17	60–63
„ —Heavy Work	13–14	55–57
Hospital Wards	18–19	65–67
Schools—Assembly Halls	13–14	55–57
„ —Classrooms	15–17	60–63
Warehouses—Storage only	10–12	50–53
—Working and Packing ..	13–14	55–57

TABLE 2

SOME RECOMMENDED NATURAL VENTILATION AIR-CHANGE RATES

Occupancy	Volumes/hr
Cinemas, Theatres	Mechanical Ventilation essential
Houses..	1–2
Factories	2–4
Schools	3
Warehouses	$\frac{1}{2}$–1
Offices	$1\frac{1}{2}$–3

201

TABLE 3

SOME COMMON CONDUCTIVITIES

Material	Density (Average) kg/m³	lb/ft³	S.I. Units 'k' $\left(\frac{Wm}{m^2{}°C}\right)$	'r' $\left(\frac{1}{k}\right)$	Imperial Units 'k' $\left(\frac{Btu\ in}{ft^2hr°F}\right)$	'r' $\left(\frac{1}{k}\right)$
Asbestos Cement Sheet ..	1520	95	0·22 to 0·29	4·55 to 3·45	1·5 to 2·0	0·67 to 0·50
,, Blanket	—	—	0·055	18·2	0·38	2·63
,, Insulating Board	—	—	0·108	9·25	0·75	1·33
,, Loose Fibre ..	—	—	0·055	18·2	0·38	2·63
,, Sprayed Fibre ..	—	—	0·046	21·8	0·32	3·12
Asphalt	2250	140	1·23	0·815	8·5	0·12
Bricks—Commons, dry ..	1870	117	1·21	0·83	8·4	0·12
,, ,, wet ..	2040	127	1·67	0·60	11·6	0·09
,, —Lightweight ..	1040	65	0·317	3·22	2·2	0·45
,, —Engineering ..	2300	144	1·15	0·87	8·0	0·12
,, —Sand–Lime ..	1920	120	1·58	0·63	11·0	0·09
,, —Diatomaceous ..	960	60	0·215	4·65	1·5	0·67
Brickwork	—	—	0·69 to 1·15	1·45 to 0·87	4·8 to 8·0	0·21 to 0·13
Concrete—1 : 2 : 4 ..	—	—	1·44	0·695	10·0	0·10
,, —Clinker	1600	100	0·33 to 0·40	3·0 to 2·5	2·3 to 2·8	0·44 to 0·36
,, —Vermiculite ..	480	30	0·10 to 0·15	10·0 to 6·65	0·70 to 1·00	1·43 to 1·00
,, ,, ..	640	40	0·19 to 0·27	5·25 to 3·70	1·30 to 1·90	0·77 to 0·53
,, —L.E.C.A. ..	720	45	0·25	4·0	1·70	0·59
,, ,, ..	960	60	0·33	3·0	2·30	0·43
,, ,, ..	1200	75	0·46	2·17	3·20	0·31
,, —Foamed Slag ..	1120	70	0·25 to 0·34	4·0 to 3·0	1·70 to 2·35	0·59 to 0·43
,, —No-fines ..	1360	85	0·46 to 0·56	2·17 to 1·78	3·20 to 3·90	0·31 to 0·26
,, ,, ..	1760	110	0·84 to 0·94	1·19 to 1·06	5·80 to 6·50	0·17 to 0·15
,, —Pumice	960	60	0·19 to 0·29	5·25 to 3·45	1·30 to 2·0	0·77 to 0·50
,, —Cellular ..	335	21	0·10	10·0	0·70	1·43
,, ,, ..	1280	80	0·58	1·72	4·0	0·25
Corkboard	—	—	0·053	18·9	0·37	2·70
Cork—Granulated ..	—	—	0·039 to 0·043	25·6 to 23·2	0·27 to 0·30	3·70 to 3·33
,, —Regranulated Baked	—	—	0·046 to 0·052	21·8 to 19·2	0·32 to 0·36	3·13 to 2·78
Eel Grass Quilt	—	—	0·036 to 0·039	27·8 to 25·6	0·25 to 0·27	4·0 to 3·70
Ebonite—Expanded ..	—	—	0·03 to 0·05	33·3 to 20·0	0·20 to 0·32	5·0 to 3·0
Felt (wool, hair or jute) ..	—	—	0·039	25·6	0·27	3·70
Felt, Bituminous	1120	70	0·20	5·0	1·4	0·71
Fibreboard—Bit.-bonded ..	370	23	0·058	17·3	0·41	2·44
,, —Bit.-impreg. ..	430	27	0·072	13·9	0·50	2·0
,, —B'lding-board	400	25	0·065	15·4	0·45	2·22
,, ,,	560	35	0·079	12·6	0·55	1·82
,, ,,	720	45	0·094	10·6	0·65	1·54
Foamed Slag—Coarse ..	480	30	0·10 to 0·13	10·0 to 7·69	0·68 to 0·90	1·47 to 1·11
,, ,, —Fine ..	670	42	0·121	8·33	0·84	1·19
Glass Wool Mats—Plain ..	24	1·5	0·042	25·0	0·29	3·45
,, ,, ,, ,,	48	3·0	0·033	33·3	0·23	4·35
,, ,, ,, —Bonded	48	3·0	0·036	25·0	0·25	4·0
Glass Sheet—Window ..	2500	157	1·05	0·91	7·30	0·14
Mineral Wool—Loose ..	—	—	0·03 to 0·04	33·3 to 25·0	0·23 to 0·30	4·1 to 3·33

TABLE 3—*continued*

Material	Density (Average)		S.I. Units		Imperial Units	
	$\dfrac{kg}{m^3}$	$\dfrac{lb}{ft^3}$	'k' $\left(\dfrac{Wm}{m^2{}^\circ C}\right)$	'r' $\left(\dfrac{1}{k}\right)$	'k' $\left(\dfrac{Btu\ in}{ft^2hr{}^\circ F}\right)$	'r' $\left(\dfrac{1}{k}\right)$
Mineral Wool—Mats ..	80	5	0·035	28·6	0·24	4·16
,, ,, —Quilts ..	128	8	0·036	27·8	0·25	4·0
,, ,, ,, ..	208	12	0·042	23·8	0·29	3·45
Plaster—Gypsum	1280	80	0·460	21·8	3·20	0·31
,, —Perlite	610	38	0·187	5·37	1·30	0·77
,, —Sand/Cement ..	1570	98	0·535	1·87	3·70	0·27
,, —Plasterboard ..	960	60	0·158	6·32	1·10	0·91
Polystyrene—Expanded ..	14	0·9	0·035	28·6	0·24	4·15
,, ,, ..	24	1·5	0·033	30·0	0·23	4·35
,, ,, ..	32	2·0	0·030	33·3	0·21	4·75
Polyurethane—Expanded ..	40	2·5	0·037	27·1	0·26	3·85
,, ,, ..	80	5·0	0·040	25·0	0·28	3·55
Stone—Granite	2650	165	2·93	0·342	20·3	0·05
,, —Limestone	2180	136	1·53	0·652	10·6	0·09
,, —Marble	2750	170	2·52	0·398	17·4	0·06
,, —Sandstone	2000	125	1·30	0·769	9·0	0·11
,, —Slate	2750	170	1·88	0·531	13·0	0·08
Strawboard	255	16	0·10	10·0	0·6	1·67
Thatch—Reed	270	17	0·086	11·6	0·6	1·67
,, —Straw	240	15	0·072	13·9	0·5	2·0
Tiles, Flooring—Clay ..	1920	120	0·84	1·19	5·8	0·17
,, ,, —Concrete..	2150	135	1·15	0·87	8·0	0·13
,, ,, —Cork ..	530	33	0·084	11·9	0·58	1·72
,, ,, —Plastic ..	1040	65	0·51	1·96	3·5	0·29
,, ,, ,, ..	1760	110	0·375	2·68	2·6	0·38
,, ,, —PVC/ asbestos	2080	130	0·865	1·16	6·0	0·17
,, ,, —Rubber ..	1600	100	0·304	3·30	2·1	0·48
Timber—Deal	620	38	0·125	8·0	0·87	1·15
,, —Oak	770	48	0·158	6·33	1·11	0·90
,, —Plywood	530	33	0·138	7·25	0·96	1·04
,, —Chipboard ..	640	40	0·122	8·2	0·85	1·18
Vermiculite—Loose ..	—	—	0·039	25·7	0·27	3·70
Water	1000	62·4	0·60	1·67	4·15	0·24
Wood-wool Slabs	400	25	0·079	12·6	0·55	1·82
,, ,, ...	800	50	0·13	7·7	0·90	1·11
Wool—Aluminium ..	29	1·8	0·043	23·2	0·30	3·33
,, ,, ..	40	2·5	0·173	5·78	1·20	0·83
,, —Steel	104	6·5	0·108	9·26	0·75	1·33

TABLE 4

RECOMMENDED AIR CHANGES (FRANCE)

(Taken from *Manuel des Industries Thermique*, published by Co.S.T.I.C.)

Occupancy	Rates of Air Changes per hour
Offices	3
Cinemas and Theatres:	
Smoking not allowed ..	3
Smoking allowed ..	5
Kitchens—Small	20
,, —Medium	15
,, —Large	10
Schools	3
Laboratories	5
Wash-houses, Laundries ..	5
Swimming Baths	3
Restaurants	4
Bathrooms	4
Operating Theatres	4
Assembly Rooms	5
Cloak Rooms	4
W.C.	5

When a location is subject to pollution by gas or inflammable vapour the rate of ventilation must be increased so as to prevent the level of concentration exceeding safe limits. In certain particular cases where calculation may be difficult the following levels should be adopted:

Garage	4 air changes per hour
Painting	5 ,, ,, ,, ,,
Spray gun painting ..	23 ,, ,, ,, ,,
Battery charging rooms ..	5 ,, ,, ,, ,,

TABLE 5

SENSIBLE HEAT AND WATER VAPOUR EMISSION BY HUMANS

(Taken from *Manuel des Industries Thermique*, published by Co.S.T.I.C.)

Activity	Total Heat Dissipated (Watts)	15°C(d.b)		18°C(d.b)		20°C(d.b)		23°C(d.b)		26°C(d.b)		29°C(d.b)	
		Sensible Heat (Watts)	Water Vapour (grammes/hr)	Sensible Heat (Watts)	Water Vapour (grammes/hr)	Sensible Heat (Watts)	Water Vapour (grammes/hr)	Sensible Heat (Watts)	Water Vapour (grammes/hr)	Sensible Heat (Watts)	Water Vapour (grammes/hr)	Sensible Heat (Watts)	Water Vapour (grammes/hr)
At rest: seated ..	112	95	26	90	33	86	40	74	58	66	70	46	98
,, ,, standing ..	126	106	31	99	42	90	54	79	72	66	91	46	122
Office Worker/light work	144	116	42	107	56	96	72	81	95	66	117	46	147
Shop Assistant: standing: quiet period ..	174	130	67	115	89	101	110	83	146	66	163	46	200
Shop Assistant: standing: busy period ..	193	135	88	119	110	108	128	85	163	66	191	46	227
Walking (at 3·2 km/h (2 m.p.h.))	223	151	109	133	135	121	154	93	196	73	226	50	260
Mechanic/House Painter	250	165	130	145	160	130	182	101	226	81	256	52	300
Waiter: busy period ..	269	181	168	158	204	140	230	111	274	95	298	70	337
Walking (at 3·8 km/h (2½ m.p.h.)) ..	307	189	179	165	216	147	242	118	286	101	312	74	353
Walking (at 6·4 km/h (4 m.p.h.))	406	238	255	203	307	180	342	151	386	133	412	102	460

N.B.—Table 5 gives mean levels for normal individuals (surface area 1·80 m² (20 ft²)): the figures are only valid for atmospheres where relative humidity is between 30% and 80%.

For wholly female occupation: Reduce the figures in the table by 20%.

For wholly infant occupation: Reduce the figures in the table by 20% to 40% according to age.

For mixed occupation:
    (*a*) If the proportions are known, calculate each separately.
    (*b*) If the proportions are not known, reduce the figures by 10%.

TABLE 6

RECOMMENDED FRESH AIR REQUIREMENTS

(Taken from *Manuel des Industries Thermique*, published by Co.S.T.I.C.)

Occupancy	Volume per occupant		Fresh Air Requirement (Air Changes/hr)
	m^3	ft^3	
Adults	2	70	15
	5	180	6
	10	360	2
	15	540	0·8
Children	3	105	15
	5	180	7·5
	10	360	2·5
	15	540	1·2
Workers	5	180	8
	10	360	3
	15	540	1·5

In an inhabited space the occupants cause the level of oxygen to decrease and the level of carbon dioxide to increase. Due to infiltration, this normally is not a serious problem, but in transportation vehicles or in well-sealed locations vitiation of the air by the occupants can arise.

Further, the problem of body odours is often important and is a fundamental element in ventilation calculations.

When there is no other cause of pollution than the occupants themselves, and when there are no gases or inflammable vapours produced, the figures given in the above table may be adopted for fresh air requirements.

(Freely translated from the original text)

TABLE 7

VAPOUR DIFFUSIVITIES

N.B.—The values given in this table are not always related to a specific type or density of the material named, and some values for apparently similar materials conflict with each other. The source of each value is noted.

Material	S.I. Units	Imperial Units		Source*
	gm mm/m² sec b	lb in/ft² h atm	gr in/ft² h mb	
Brick	4·00	0·12	—	(a)
,, ...	6·25	—	0·128	(b)
,, ...	6·35	—	0·13	(c)
Insulating Board ..	1·65 to 6·70	0·05 to 0·20	—	(a)
,, ,, ..	2·23 to 7·25	—	0·45 to 1·50	(b)
,, ,, ..	2·18	—	0·44	(c)
Concrete	0·99 to 3·40	0·03 to 0·10	—	(a)
Timber	1·32 to 2·00	0·04 to 0·06	—	(a)
,, Spruce	1·35	—	0·28	(b)
,, Pine	0·12 to 0·22	—	0·025 to 0·045	(b)
Plywood	0·0165 to 0·0665	0·0005 to 0·002	—	(a)
,, ...	0·195	—	0·04	(b)
Hardboard	0·132 to 0·20	0·004 to 0·006	—	(a)
Plasterboard	2·00	0·06	—	(a)
,, ...	1·90	—	0·39	(b)
Plaster	1·56 to 2·58	—	0·32 to 0·52	(b)
Compressed Strawboard	0·67 to 1·32	0·02 to 0·04	—	(a)
Expanded Polystyrene ..	0·169 to 0·67	0·005 to 0·02	—	(a)
Foamed Urea-formalde- hyde	3·02 to 5·00	0·09 to 0·15	—	(a)
Foamed Polyurethane ..	0·099 to 0·337	0·003 to 0·10	—	(a)
Expanded Ebonite ..	0·00165 to 0·0082	0·00005 to 0·00025	—	(a)

Vapour Diffusances for Thin Film Materials				
Material	gm/m² sec b	lb/ft² h atm	gr/ft² h mb	Source
Gloss Paint	0·0027 to 0·013	0·002 to 0·01	—	(a)
Polythene Sheet	0·0004	0·0003	—	(a)
Aluminium Foil	0·00003	0·00002	—	(a)

* *Source:* (a) Ball; (b) Billington; (c) Shaw

Appendix C: Conversion Charts

On the four following pages, two conversion charts (Chart A, pages 210-211 and Chart B, *pages 212-213) show the Imperial Units' equivalents of Système International Units, for various requirements.*

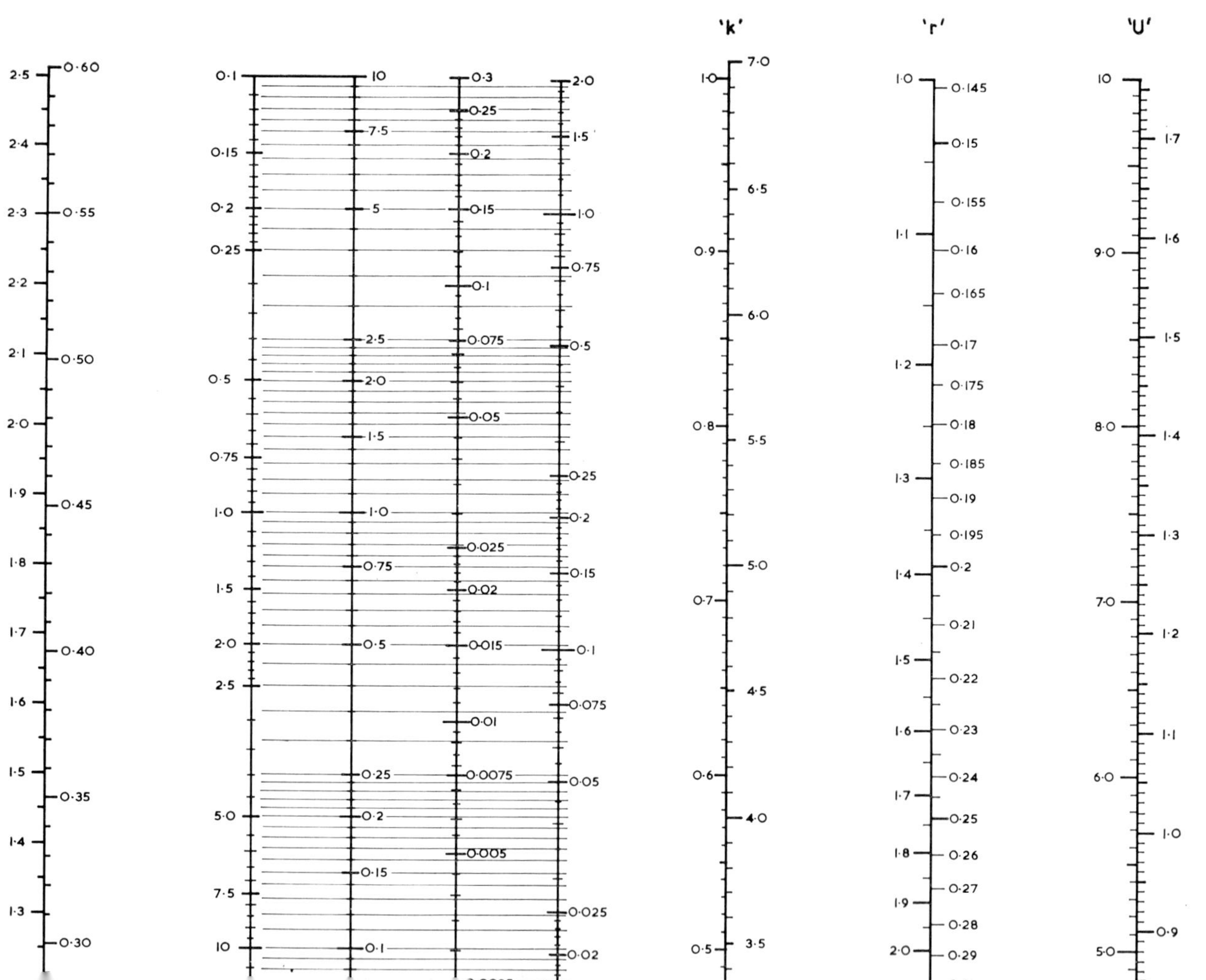

CHART A
Imperial
Equivalents
of S.I.
Units
'k'
'r'
'U'

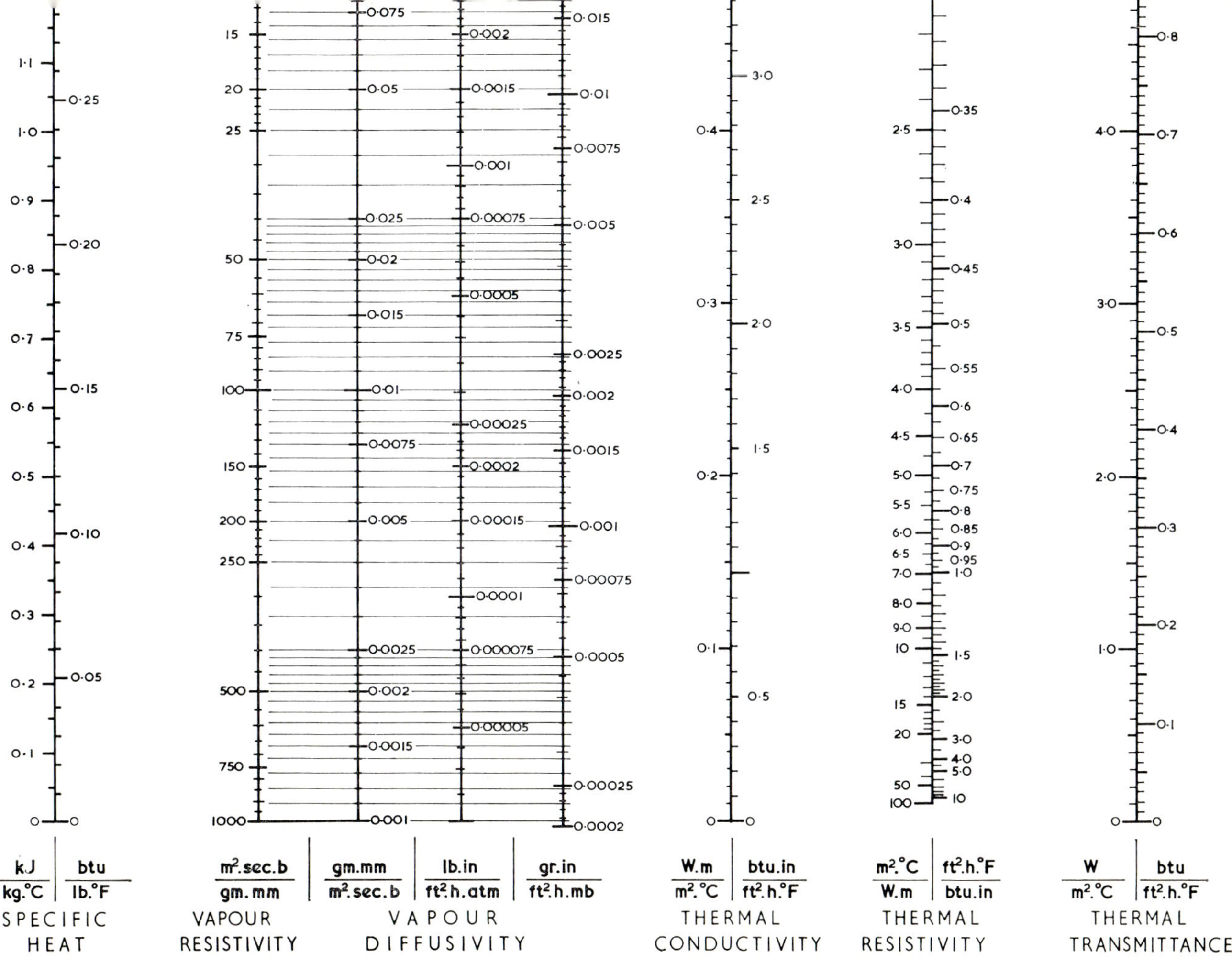

SPECIFIC HEAT
kJ / kg.°C
btu / lb.°F
VAPOUR RESISTIVITY
m².sec.b / gm.mm
VAPOUR DIFFUSIVITY
gm.mm / m².sec.b
lb.in / ft².h.atm
gr.in / ft².h.mb
THERMAL CONDUCTIVITY
W.m / m².°C
btu.in / ft².h.°F
THERMAL RESISTIVITY
m².°C / W.m
ft².h.°F / btu.in
THERMAL TRANSMITTANCE
W / m².°C
btu / ft².h.°F

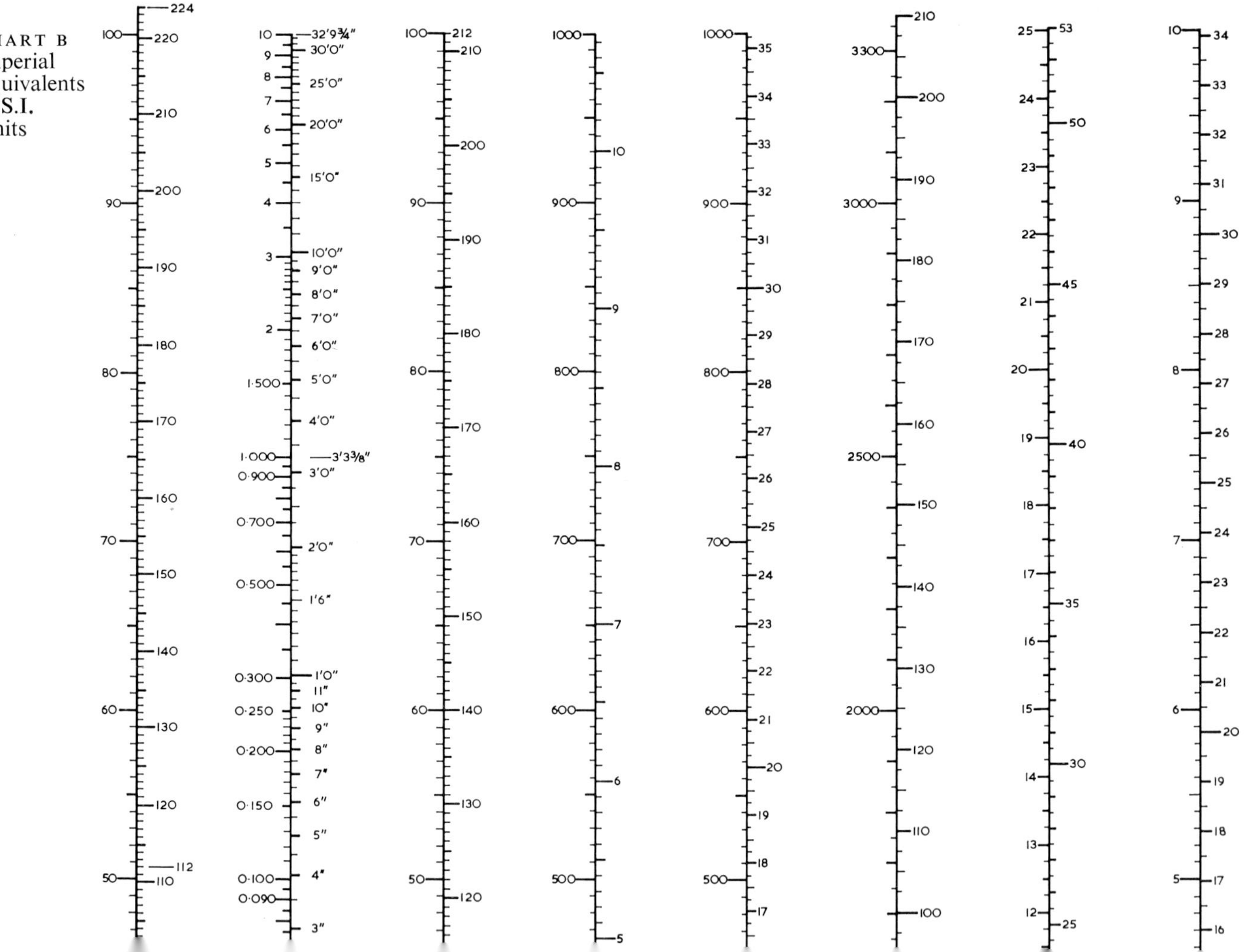
CHART B
Imperial
Equivalents
of S.I.
Units

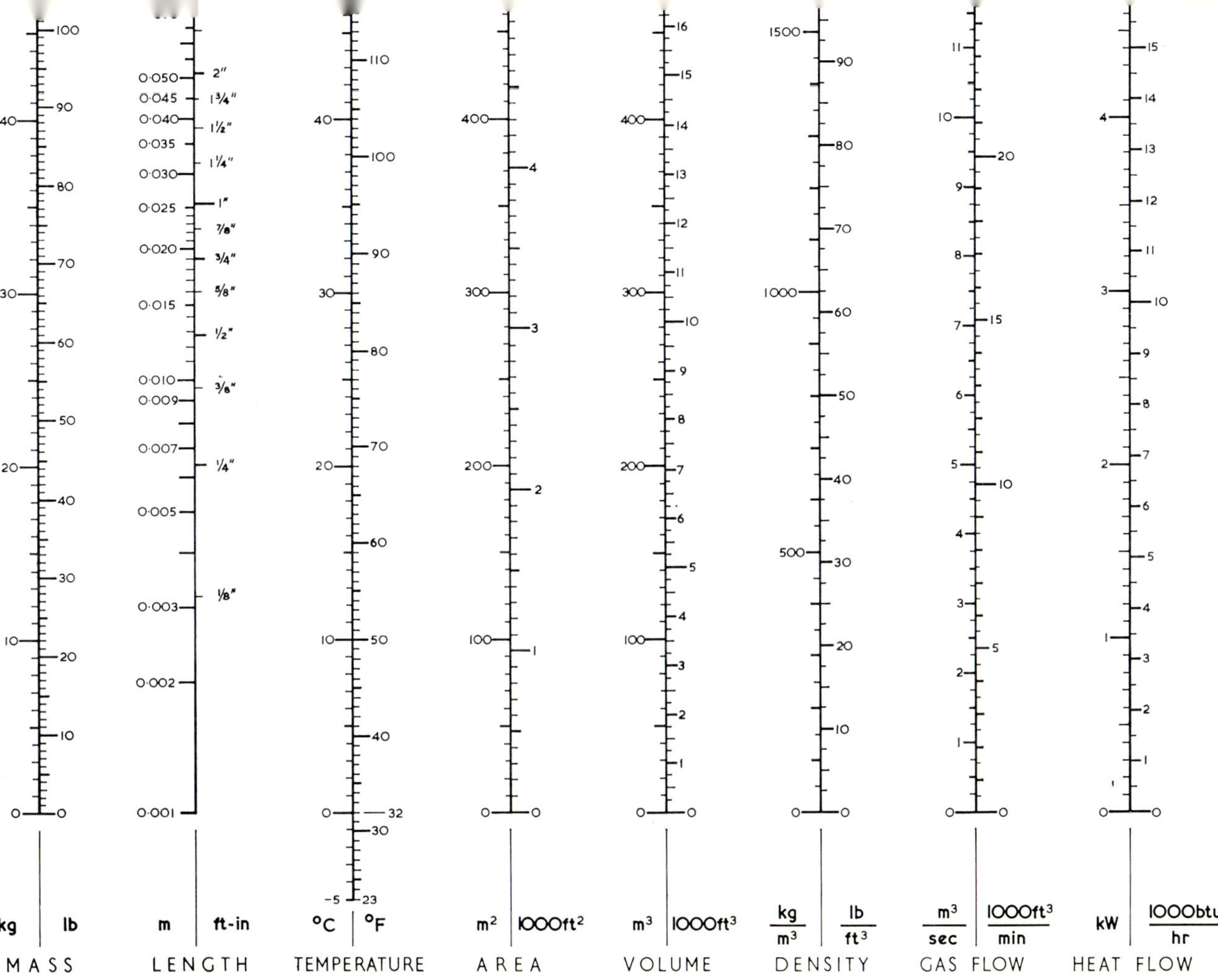

kg | lb
MASS
m | ft-in
LENGTH
°C | °F
TEMPERATURE
m² | 1000ft²
AREA
m³ | 1000ft³
VOLUME
kg/m³ | lb/ft³
DENSITY
m³/sec | 1000ft³/min
GAS FLOW
kW | 1000btu/hr
HEAT FLOW

Abbreviations

THE FOLLOWING are the principal abbreviations used in this book:

atm	=standard atmosphere
b	=bar (1000 mb)
Btu	=British thermal units
C	=thermal conductance ('k' × thickness of material)
Cp	=specific heat
d	=vapour diffusivity
D.B.	=dry-bulb temperature
d.p.	=dewpoint
H.P.H.W.	=high-pressure hot water
k	=thermal conductivity
L.P.H.W.	=low-pressure hot water
mb	=millibar (bar ÷ 1000)
M.P.H.W.	=medium-pressure hot water
M.R.T.	=mean radiant temperature
r	=thermal resistivity
R.H.	=relative humidity
Rsi	=surface resistance (inside)
Rso	=surface resistance (outside)
R_T	=thermal resistance
td	=temperature drop
t_i	=inside temperature
t_o	=outside temperature
Tsi	=inside surface temperature
Tso	=outside surface temperature
U	=thermal transmittance through a material
vr	=vapour resistivity
vR	=vapour resistance
W.B.	=wet bulb temperature

Bibliography

(Suggested publications for further reading)

ENVIRONMENTAL WARMTH AND ITS MEASUREMENT, by T. Bedford. (Medical Research Council, No. 17: H.M.S.O.)

H.E.V.A.C. PROCEEDINGS OF THE INTERNATIONAL CONFERENCE, 1961. (Institution of Heating and Ventilating Engineers)

AIR-CONDITIONING OF LARGE BUILDINGS, by K. Fowler. (The Institute of Refrigeration Engineers: Session 1962–63)

HEAT LOSS CALCULATIONS AND CALCULATION OF FUEL CONSUMPTION. (SfBA68 in *The Architects' Journal*)

HEATING INSTALLATIONS AND EQUIPMENT. (SfB56 in *The Architects' Journal*)

THERMAL DISCOMFORT IN AN EQUATORIAL CLIMATE, by G. C. Webb. (*I.H.V.E. Journal*, January, 1960)

HEATING AND AIR-CONDITIONING OF BUILDINGS, by Oscar Faber and J. R. Kell. (1966: The Architectural Press Ltd.)

METEOROLOGICAL DATA AND DESIGN TEMPERATURES, by H. C. Jamieson. (*I.H.V.E. Journal*, 1954)

AIR-CONDITIONING IN MODERN STRUCTURES, by F. S. Snow and W. L. Swain. (Institution of Civil Engineers: Paper 6676/1963)

BASIC PRINCIPLES OF VENTILATION AND HEATING, by T. Bedford. (H. K. Lewis)

I.H.V.E. GUIDE, 1965. (The Institution of Heating and Ventilating Engineers

A.S.H.R.A.E. GUIDE AND DATA BOOK. (American Society Heating, Refrigeration and Air-Conditioning)

METEOROLOGICAL GLOSSARY (M.O.729), by D. H. McIntosh (H.M.S.O.)

HEATING AND HOT-WATER SERVICES, by E. W. Shaw. (Crosby Lockwood and Son Ltd.)

HOMES FOR TODAY AND TOMORROW (Parker Morris Report). M.O.H.L.G. (H.M.S.O.)

MECHANICAL VENTILATION OF INTERIOR ROOMS. Design Bulletin 3, Part III. (M.O.H.L.G.)

THERMAL INSULATION, C.R.B.4, by N. S. Billington. (T.D.A. Ltd.)

CONSTRUCTION AND FINISH OF FLOORS FOR ELECTRICAL WARMING. (Electrical Development Association)

CONSTRUCTION AND FINISHES OF FLOORS, ETC., FOR RADIANT HEATING. (Panel Warming Association)

INSOLATION AND FENESTRATION, by A. C. Hardy and P. E. O'Sullivan. (1967: Oriel Press)

BUILDING RESEARCH STATION PUBLICATIONS, by The Building Research Station Department of Scientific and Industrial Research. (H.M.S.O.):

Digest Series No. 1
18   *Smoky Chimneys*
23   *Condensation Problems*
34   *Natural Ventilation*
35   *Heat Loss from Dwellings*
58   *Condensation in Walls*
94   *Domestic Heating*
132   *Condensation*
133   *Domestic Heating and Thermal Insulation*

Design Series
7   *Air Conditioning Methods and Equipment in Temperate Climates*, by D. E. Sexton
24   *Condensation in Large Panel Construction*, by E. F. Ball
26   *Heating for High-Density Housing*, by E. A. Milroy and G. A. Atkinson
35   *Frameless Large Panel Systems of Construction*, by A. Alsop and R. B. Bonshor
38   *Studies of Air Flow round Buildings*, by A. F. E. Wise, D. E. Sexton and M. S. T. Lillywhite

Research Series
   37   *Investigation of Summer Over-heating, England*, by A. G. Louden and E. Danter
   39   *Sunpath Diagrams and Overlays for Solar Heat Gain Calculations*, by P. Petheridge

Engineering Series
   17   *Technical Aspects of Thermal Insulation in Buildings*, by A. W. Pratt

BRITISH STANDARDS INSTITUTION:
Metric Conversion and Standards. P.D.5686. (B.S.I.: H.M.S.O.)
The International System (S.I.) Units. B.S.3763 and B.S.874 (1965). (B.S.I.: H.M.S.O.)
The Thermal Insulation of Buildings, by G. D. Nash, J. Comrie and H. F. Broughton. (1955: H.M.S.O.)

POST-WAR BUILDING STUDIES:
   19   *Heating and Ventilation of Dwellings.* (H.M.S.O.)
   27   *Heating and Ventilation of Schools.* (H.M.S.O.)
   32   *District Heating: Part IV.* (H.M.S.O.)

Index